Access Your Online Resources

Don't miss out on the Online Resources included with your purchase!

Your purchase of this product unlocks access to our Online Resources page. Elevate your study experience with our **interactive practice test interface**, along with all of the additional resources that we couldn't include in this book.

Flip to the Online Resources section at the end of this book to find the link and a QR code to get started!

Mometrix
TEST PREPARATION

Mometrix
TEST PREPARATION

Ham Radio

Technician License Book

Ham Radio Secrets Study Guide for the ARRL® Technician License Exam

2026 Question Pool

Written and edited by Matthew Bowling

Printed in the United States of America

This paper meets the requirements of ANSI/NISO Z39.48-1992 (Permanence of Paper).

Paperback
ISBN 13: 978-1-5167-3373-6
ISBN 10: 1-5167-3373-8

Dear Future Exam Success Story

First of all, **THANK YOU** for purchasing Mometrix study materials!

Second, congratulations! You are one of the few determined test-takers who are committed to doing whatever it takes to excel on your exam. **You have come to the right place.** We developed these study materials with one goal in mind: to deliver you the information you need in a format that's concise and easy to use.

In addition to optimizing your guide for the content of the test, we've outlined our recommended steps for breaking down the preparation process into small, attainable goals so you can make sure you stay on track.

We've also analyzed the entire test-taking process, identifying the most common pitfalls and showing how you can overcome them and be ready for any curveball the test throws you.

Standardized testing is one of the biggest obstacles on your road to success, which only increases the importance of doing well in the high-pressure, high-stakes environment of test day. Your results on this test could have a significant impact on your future, and this guide provides the information and practical advice to help you achieve your full potential on test day.

Your success is our success

We would love to hear from you! If you would like to share the story of your exam success or if you have any questions or comments in regard to our products, please contact us at **800-673-8175** or **support@mometrix.com**.

Thanks again for your business and we wish you continued success!

Sincerely,
The Mometrix Test Preparation Team

Need more help? Check out our flashcards at:
http://MometrixFlashcards.com/hamradio

Written and edited by the Mometrix Exam Secrets Test Prep Team
Printed in the United States of America

TABLE OF CONTENTS

Introduction

The FCC (Federal Communications Commission) has **three levels of licensure**: Technician class, General class, and Extra class. Each class provides further permissions. The Technician class allows for basic amateur radio use.

To aid in the process of studying, the FCC releases the full question pool for the exam. The question pool for the Technician class exam is made up of **35 topics**. Each test will contain one question from each topic. The question pool contains enough questions to provide for at least 10 unique tests.

Because this guide contains the full question pool and 10 practice tests, it is possible to familiarize yourself with the questions to a point where you can pass the test on rote memory alone. However, to do so would be to miss out on a great opportunity to discover what knowledge gaps you have in the field of amateur radio and address them. We recommend that you **study for memory and comprehension simultaneously**. One possible strategy is outlined below:

1. Read through the question bank once to understand the full scope of the topics.
2. Address any obvious knowledge gaps with supplemental research.
3. Take a practice test and score yourself.
4. Examine each question you answered incorrectly and research its topic until you feel confident with that content.
5. Reread the question bank.
6. Repeat steps 3-5 until you comprehend each topic and its questions.

Of course, you should optimize any study plan for your individual strengths. No matter what study plan you follow, if you study diligently, you can be prepared and confident on your test day.

Technician Question Pool

T1 – FCC Rules

T1A – Purpose and permissible use of the Amateur Radio Service; Operator/primary station license grant; Meanings of basic terms used in FCC rules; Interference; RACES rules; Phonetics; Frequency Coordinator; Beacon

T1A01. Which of the following is part of the Basis and Purpose of the Amateur Radio Service?

A. Providing personal radio communications for as many citizens as possible
B. Providing communications for international contesting
C. Advancing skills in the technical and communication phases of the radio art
D. All these choices are correct

C: As stated in the FCC rules, advancing skills in the technical and communication phases of the radio art is a purpose of the Amateur Radio Service

T1A02. Which agency regulates and enforces the rules for the Amateur Radio Service in the United States?

A. ARRL
B. Homeland Security
C. The FCC
D. All these choices are correct

C: The Federal Communications Commission (FCC) is the agency that regulates and enforces the rules for the Amateur Radio Service in the United States.

T1A03. What do the FCC rules state regarding the use of a phonetic alphabet for station identification in the Amateur Radio Service?

A. It is required when transmitting emergency messages
B. It is encouraged when using phone emissions
C. It is required when in contact with foreign stations
D. All these choices are correct

B: By a phone emission in the English language. Use of a phonetic alphabet as an aid for correct station identification is encouraged.

T1A04. How do you receive official notification of your new license and call sign after passing the exam?

A. Email from the FCC with a link to download the license grant
B. USPS First-Class Mail from the FCC
C. Email from the lead volunteer examiner
D. USPS First-Class Mail from the lead volunteer examiner

A: Official notification of new licensure and call signs after passing the exam are received by email from the FCC with a link to download the license grant.

T1A05. What proves that the FCC has issued an operator/primary license grant?

A. A printed copy of the certificate of successful completion of examination
B. An email notification from the NCVEC granting the license
C. The license appears in the FCC ULS database
D. All these choices are correct

C: When transmitting, each amateur station must have a control operator. The control operator must be a person:

(a) For whom an amateur operator/primary station license grant appears on the ULS consolidated licensee database, or

(b) Who is authorized for alien reciprocal operation by § 97.107 of this part.

T1A06. On which of the following HF frequencies can automatically controlled amateur propagation beacons be found?

A. Anywhere in the Technician portion of 10 meters
B. Anywhere, if they are registered with the FCC
C. In a specific portion of each of the HF Technician bands
D. On ten meters, between 28.200 MHz and 28.300 MHz

D: Automatically controlled amateur propagation beacons can be found on ten meters between 28.200 MHz and 28.300 MHz.

T1A07. What is the FCC Part 97 definition of a space station?

A. Any satellite orbiting Earth
B. A manned satellite orbiting Earth
C. An amateur station located more than 50 km above Earth's surface
D. An amateur station using amateur radio satellites for relay of signals

C: A space station is an amateur station located more than 50 km above the Earth's surface.

T1A08. Who recommends transmit/receive channels for repeater and auxiliary stations?

A. A Frequency Spectrum Manager appointed by the FCC
B. A Volunteer Frequency Coordinator recognized by local amateurs
C. An FCC Regional Field Office
D. The International Telecommunication Union

B: A frequency coordinator recommends transmit/receive channels and other parameters for auxiliary and repeater stations. A frequency coordinator is a person or group of people who advise users of the appropriate repeater frequencies, with the intention of diminishing interference between repeaters who are operating in roughly the same frequency in the same geographical area.

T1A09. Who selects a Frequency Coordinator?

A. The FCC Office of Spectrum Management and Coordination Policy
B. The local chapter of the Office of National Council of Independent Frequency Coordinators
C. Amateur operators in a local or regional area whose stations are eligible to be repeater or auxiliary stations
D. The FCC Regional Field Office

C: Amateur operators in a local or regional area whose stations are eligible to be auxiliary or repeater stations may select a frequency coordinator. [97.3(a)(22)]

T1A10. Besides an FCC-issued amateur operator license, what is required to be the control operator of a Radio Amateur Civil Emergency Service (RACES) Station?

A. A written recommendation by the local ARRL Emergency Coordinator
B. Membership in the Amateur Radio Emergency Service (ARES)
C. Certification of current enrollment by a civil defense organization
D. Nothing

C: Certification of current enrollment by a civil defense organization is required, in addition to an FCC-issued amateur operator license, to be the control operator of a RACES station. [97.407(a)]

T1A11. Which of the following is prohibited?

A. International communications on VHF bands and higher frequencies
B. Willful or malicious interference
C. Third-party traffic using digital modes
D. All these choices are correct

B: Willful or malicious interference is prohibited.

T1B – Frequency allocations; Emission modes; Spectrum sharing; Transmissions near band edges; Contacting the International Space Station; Power output

T1B01. Which of the following frequency ranges are available for phone operation by Technician licensees?

A. 28.050 MHz to 28.150 MHz
B. 28.100 MHz to 28.300 MHz
C. 28.300 MHz to 28.500 MHz
D. 28.500 MHz to 28.600 MHz

C: Frequencies in the 28.300 MHz to 28.500 MHz range are available for phone operation by Technician licensees. [97.301(e)]

T1B02. Which of the following U.S. amateur radio operators are allowed to contact the International Space Station (ISS) on VHF bands?

A. Only amateurs with a General class or higher license
B. Any amateur with a Technician class or higher license
C. Only amateurs with a General class or higher license, and NASA approval
D. Any amateurs with a Technician class or higher license, and NASA approval

B: For a station having a control operator who has been granted a Technician, General, Advanced, or Amateur Extra Class operator license or who holds a CEPT radio-amateur license or IARP of any class. The following frequency bands and segments are authorized to space stations:

(1) The 17 m, 15 m, 12 m, and 10 m bands, 6 mm, 4 mm, 2 mm and 1 mm bands; and

(2) The 7.0-7.1 MHz, 14.00-14.25 MHz, 144-146 MHz, 435-438 MHz, 2400-2450 MHz, 3.40-3.41 GHz, 5.83-5.85 GHz, 10.45-10.50 GHz, and 24.00-24.05 GHz segments.

[97.301, 97.207(c)]

T1B03. Which frequency is in the 6-meter amateur band?

A. 49.00 MHz
B. 52.525 MHz
C. 28.50 MHz
D. 222.15 MHz

B: 52.525 MHz is within the 6-meter band. [97.301(a)]

T1B04. Which amateur band includes 146.52 MHz?

A. 6 meters
B. 20 meters
C. 70 centimeters
D. 2 meters

D: If your station is transmitting on 146.52 MHz, it is using the 2-meter amateur band. [97.301(a)]

T1B05. Which of the following bands include frequencies where Technicians are authorized to use digital modes such as FT8?

A. 10 meters
B. 6 meters
C. 2 meters
D. All these choices are correct

D: All of the provided choices are correct. [97.301, 97.305]

T1B06. On which HF bands does a Technician class operator have phone privileges?

A. None
B. 10-meter band only
C. 80-meter, 40-meter, 15-meter, and 10-meter bands
D. 30-meter band only

B: A Technician Class control operator can use 80, 40, 15, and 10m wavelengths, but only the 10m for phone privileges. [97.301(e), 97.305]

T1B07. Which of the following VHF/UHF band segments are limited to CW only?

A. 50.0 MHz to 50.1 MHz and 144.0 MHz to 144.1 MHz
B. 219 MHz to 220 MHz and 420.0 MHz to 420.1 MHz
C. 902.0 MHz to 902.1 MHz
D. All these choices are correct

A: 50.0 MHz to 50.1 MHz and 144.0 MHz to 144.1 MHz are the only ranges in VHF/UHF that are for CW only. [97.305(a), (c)]

T1B08. How are US amateurs restricted in segments of bands where the Amateur Radio Service is secondary?

A. U.S. amateurs may find non-amateur stations in those segments, and must avoid interfering with them
B. U.S. amateurs must give foreign amateur stations priority in those segments
C. International communications are not permitted in those segments
D. Digital transmissions are not permitted in those segments

A: The amateur service is secondary in some portions of the 70 cm band because U.S. amateurs may find non-amateur stations in the bands and must avoid interfering with them. [97.303]

T1B09. Why should you not set your transmit frequency to be exactly at the edge of an amateur band or sub-band?

A. To allow for calibration error in the transmitter frequency display
B. So that modulation sidebands do not extend beyond the band edge
C. To allow for transmitter frequency drift
D. All these choices are correct

D: There are several reasons why an operator should not set the transmit frequency to be exactly at the edge of an amateur band or sub-band. To begin with, this allows for calibration error in the transmitter frequency display. It also ensures that modulation sidebands do not extend beyond the band edge. Finally, it allows for transmitter frequency drift. [97.101(a), 97.301(a-e)]

T1B10. Where may SSB phone be used in amateur bands above 50 MHz?

A. Only in sub-bands allocated to General class or higher licensees
B. Only on repeaters
C. In at least some segment of all these bands
D. On any band if the power is limited to 25 watts

C: There is some part of all the amateur bands above 50 MHz that is available for SSB. [97.305(c)]

T1B11. What is the maximum peak envelope power output for Technician class operators in their HF band segments?

A. 200 watts
B. 100 watts
C. 50 watts
D. 10 watts

A: No station may transmit with a transmitter power output exceeding 200 W PEP:

(1) On the 10.10-10.15 MHz segment;

(2) On the 3.525-3.60 MHz, 7.025-7.125 MHz, 21.025-21.20 MHz, and 28.0-28.5 MHz segment when the control operator is a Novice Class operator or a Technician Class operator; or

(3) The 7.050-7.075 MHz segment when the station is within ITU Regions 1 or 3. [97.313]

T1B12. Except for some specific restrictions, what is the maximum peak envelope power output for Technician class operators using frequencies above 30 MHz?

A. 50 watts
B. 100 watts
C. 500 watts
D. 1500 watts

D: No station may transmit with a transmitter power exceeding 1.5 kW PEP. [97.313(b)]

T1C – Licensing: classes, sequential and vanity call sign systems, places where the Amateur Radio Service is regulated by the FCC, name and address on FCC license database, term, renewal, grace period, maintaining mailing address; International communications

T1C01. For which classes of amateur radio licenses does the FCC currently issue new licenses?

A. Novice, Technician, General, Amateur Extra
B. Technician, Technician Plus, General, Amateur Extra
C. Novice, Technician Plus, General, Advanced
D. Technician, General, Amateur Extra

D: The classes of amateur operator license grants are: Technician, General, and Amateur Extra. The person named in the operator license grant is authorized to be the control operator of an amateur station with the privileges authorized to the operator class specified on the license grant. [97.9(a), 97.17(a)]

T1C02. Who may select a desired call sign under the vanity call sign rules?

A. Only a licensed amateur with a General or Amateur Extra Class license
B. Only a licensed amateur with an Amateur Extra Class license
C. Only a licensed amateur who has been licensed continuously for more than 10 years
D. Any licensed amateur

D: The person named in an operator/primary station license grant or in a club station license grant is eligible to make application for modification of the license grant, or the renewal thereof, to show a call sign selected by the vanity call sign system. [97.19]

T1C03. What types of international communications are FCC-licensed amateur radio stations permitted to make?

A. Communications incidental to the purposes of the Amateur Radio Service and remarks of a personal character
B. Communications incidental to conducting business or remarks of a personal nature
C. Only communications incidental to contest exchanges; all other communications are prohibited
D. Any communications that would be permitted by an international broadcast station

A: FCC-licensed amateur stations are permitted to make international communications incidental to the purposes of the amateur service and remarks of a personal character. [97.117]

T1C04. What may happen if the FCC is unable to reach you by email?

A. Fine and suspension of operator license
B. Revocation of the station license or suspension of the operator license
C. Revocation of access to the license record in the FCC system
D. Nothing; there is no such requirement

B: If the FCC is unable to reach you by email, then your station license may be revoked or your operator license may be suspended. [97.23]

T1C05. Which of the following is a valid Group D call sign format for Technician class?

A. KF1XXX
B. KA1X
C. W1XX
D. All these choices are correct

A: Of the choices listed, only the KF1XXX series of call signs is available to technician class amateur operators. The other two series listed are available only to extra class amateur operators.

T1C06. Which of the following statements is true about using your amateur radio license when operating aboard a vessel or craft in international waters?

A. Amateur operation is prohibited in international waters
B. You may only operate during maritime emergencies
C. You need special FCC authorization for maritime mobile operation
D. You may operate from a US-documented vessel with the master's permission

D: You may operate in international waters from a US-documented vessel with the master's permission. [97.5(a)(2), 97.11(a)]

T1C07. How long before the expiration date may an amateur radio license renewal be requested?

A. 30 days
B. 90 days
C. 6 months
D. 1 year

B: An amateur radio license renewal may be requested 90 days before the expiration date. [97.21, 1.949]

T1C08. What is the normal term for an FCC-issued amateur radio license?

A. Five years
B. Eight years
C. Ten years
D. Life

C: The normal term for an FCC-issued primary station/operator license grant is 10 years. [97.25]

T1C09. What is the grace period for renewal if an amateur license expires?

A. Two years
B. Three years
C. Five years
D. Ten years

A: The grace period following the expiration of an amateur license, within which the license may be renewed, is two years. [97.21(a)(b)]

T1C10. How soon after passing the examination for your first amateur radio license may you transmit on the amateur radio bands?

A. Immediately on receiving your Certificate of Successful Completion of Examination (CSCE)
B. As soon as your operator/station license grant appears on the ARRL website
C. As soon as your operator/station license grant appears in the FCC's license database
D. As soon as you receive your license in the mail from the FCC

C: After you pass the examination required for your first amateur radio license, you may operate a transmitter on an amateur service frequency as soon as your name and call sign appear in the FCC's ULS database. [97.5a]

T1C11. If your license has expired and is still within the allowable grace period, may you continue to transmit on the amateur radio bands?

A. Yes, for up to two years
B. Yes, as soon as you apply for renewal
C. Yes, for up to one year
D. No, you must wait until the license has been renewed

D: An operator whose license has expired but who is still within the allowable grace period may not continue to operate a transmitter on amateur service frequencies until the ULS database shows that the license has been renewed. [97.21(b)]

T1D – Authorized and Prohibited Transmissions; Sale of Equipment

T1D01. With which countries are FCC-licensed amateur radio stations prohibited from exchanging communications?

A. Any country whose administration has notified the International Telecommunication Union (ITU) that it objects to such communications
B. Any country whose administration has notified the American Radio Relay League (ARRL) that it objects to such communications
C. Any country banned from such communications by the International Amateur Radio Union (IARU)
D. Any country banned from making such communications by the American Radio Relay League (ARRL)

A: FCC-licensed amateur stations are prohibited from exchanging communications with any country whose administration has notified the ITU that it objects to such communications. [97.111(a)(1)]

T1D02. Under which of the following circumstances are one-way transmissions by an amateur station prohibited?

A. Announcements of upcoming ham radio operating events
B. Broadcasting
C. International Morse Code Practice
D. Telecommand or transmissions of telemetry

B: While an amateur station is broadcasting, one-way transmissions are prohibited. [97.113(b), 97.111(b)]

T1D03. When is it permissible to transmit messages encoded to obscure their meaning?

A. Only when using a remote station during a contest
B. Only when transmitting certain approved digital codes
C. Only when transmitting control commands to space stations or model craft
D. Never

C: An amateur station is allowed to hide the meaning of a transmitted message with codes or ciphers only when transmitting control commands to space stations or radio control craft. [97.211(b), 97.215(b), 97.113(a)(4)]

T1D04. Under what conditions is an amateur station authorized to transmit music using a phone emission?

A. When incidental to an authorized retransmission of manned spacecraft communications
B. When the music produces no spurious emissions
C. When transmissions are limited to less than three minutes per hour
D. When the music is transmitted above 1280 MHz

A: The only time an amateur station is authorized to transmit music is when it is incidental to authorized retransmission of manned spacecraft communications. [97.113(a)(4), 97.113(c)]

T1D05. When may amateur radio operators use their stations to notify other amateurs of the availability of equipment for sale or trade?

A. Never
B. When the equipment is not the personal property of either the station licensee, or the control operator, or their close relatives
C. When no profit is made on the sale
D. When selling amateur radio equipment and not on a regular basis

D: Amateur radio operators may use their stations to notify other amateurs of the availability of equipment for sale or trade when the equipment is normally used in an amateur station and such activity is not conducted on a regular basis. [97.113(a)(3)(ii)]

T1D06. What, if any, are the restrictions concerning transmission of language that may be considered indecent or obscene?

A. The FCC maintains a list of words that are not permitted to be used on amateur frequencies
B. Any such language is prohibited
C. The International Telecommunications Union (ITU) maintains a list of words that are not permitted to be used on amateur frequencies
D. There is no such prohibition

B: As stated by the FCC, no amateur station shall transmit obscene or indecent words or language. [97.113(a)(4)]

T1D07. Which of the following is an example of an auxiliary station?

A. A station sending one-way transmissions between a remote repeater receiver and the main repeater transmitter
B. A backup radio for emergency use in case the main station radio fails
C. A station used in the Military Auxiliary Radio System for linking government and amateur stations in a digital network
D. A second station utilized in multi-operator contest stations

A: A station sending one-way transmissions between a remote repeater receiver and the main repeater transmitter is an example of an auxiliary station. [97.113(d), 97.201(e)]

T1D08. In which of the following circumstances may the control operator of an amateur station receive compensation for operating that station?

A. When the communication is related to the sale of amateur equipment by the control operator's employer
B. When the communication is a part of classroom instruction at an educational institution
C. When the communication is made to obtain emergency information for a local broadcast station
D. All these choices are correct

B: The control operator of an amateur station may receive compensation for operating the station when the communication is incidental to classroom instruction at an educational institution. [97.113(a)(3)(iii)]

T1D09. When may amateur stations transmit information in support of broadcasting, program production, or news gathering, assuming no other means is available?

A. When such communications are directly related to the immediate safety of human life or protection of property
B. When broadcasting communications to or from the space shuttle
C. Where non-commercial programming is gathered and supplied exclusively to the National Public Radio network
D. Never

A: Assuming no other means is available, amateur stations are authorized to transmit signals related to broadcasting, program production, or news gathering only where such communications directly relate to the immediate safety of human life or protection of property. [97.113(5)(b)]

T1D10. How does the FCC define broadcasting for the Amateur Radio Service?

A. Two-way transmissions by amateur stations
B. Any transmission made by the licensed station
C. Transmission of messages directed only to amateur operators
D. Transmissions intended for reception by the general public

D: In the FCC rules for the amateur services, broadcasting refers to transmissions intended for reception by the general public. Broadcasting refers to transmissions that are meant to be received by the general public, either on relay or directly. [97.3(a)(10)]

T1D11. When may an amateur station transmit without identifying on the air?

A. When the transmissions are of a brief nature to make station adjustments
B. When the transmissions are unmodulated
C. When the transmitted power level is below 0.1 watt
D. When transmitting signals to control model craft

D: The station identification procedure is not required for an amateur station's transmissions directed only to the model craft, provided that a label indicating the station call sign and the station licensee's name and address is affixed to the station transmitter. [97.215]

T1D12. Which of the following is required when making on-the-air test transmissions?

A. Identify the transmitting station
B. Conduct tests only between 10 p.m. and 6 a.m. local time
C. Notify the FCC of the transmissions
D. All these choices are correct

A: Identify the transmitting station when making on-the-air test transmissions. [97.119(a)]

T1E – Control operator: eligibility, designating, privileges, duties, location, required; Control point; Control types: automatic, remote

T1E01. When may an amateur station transmit without a control operator?

A. When using automatic control, such as in the case of a repeater
B. When the station licensee is away and another licensed amateur is using the station
C. When the transmitting station is an auxiliary station
D. Never

D: When transmitting, each amateur station must have a control operator. [97.7(a)]

T1E02. Who may be the control operator of a station communicating through an amateur satellite or space station?

A. Only an Amateur Extra Class operator
B. A General class or higher licensee with a satellite operator certification
C. Only an Amateur Extra Class operator who is also an AMSAT member
D. Any amateur allowed to transmit on the satellite uplink frequency

D: Any amateur station may be a space station. A holder of any class operator license may be the control operator of a space station, subject to the privileges of the class of operator license held by the control operator. [97.301, 97.207(c)]

T1E03. Who must designate the station control operator?

A. The station licensee
B. The FCC
C. The frequency coordinator
D. Any licensed operator

A: The station licensee must designate the station control operator. [97.103(b)]

T1E04. What determines the transmitting frequency privileges of an amateur station?

A. The frequency authorized by the frequency coordinator
B. The frequencies printed on the license grant
C. The highest class of operator license held by anyone on the premises
D. The class of operator license held by the control operator

D: The transmitting privileges of an amateur station are determined by the class of operator license held by the control operator. [97.103(b)]

T1E05. What is an amateur station's control point?

A. The location of the station's transmitting antenna
B. The location of the station's transmitting apparatus
C. The location at which the control operator function is performed
D. The mailing address of the station licensee

C: An amateur station control point is the location at which the control operator function is performed. The control point is the set of locations at which the control operator functions are carried out. [97.3(a)(14)]

T1E06. When, other than during an emergency, may a Technician class licensee be the control operator of a station operating in an Amateur Extra Class band segment?

A. At no time
B. When designated as the control operator by an Amateur Extra Class licensee
C. As part of a multi-operator contest team
D. When using a club station whose trustee holds an Amateur Extra Class license

A: Each class of license allows the licensee to operate only within the allotted frequencies at their level or below. [97.301]

T1E07. When the control operator is not the station licensee, who is responsible for the proper operation of the station?

A. All licensed amateurs who are present at the operation
B. Only the station licensee
C. Only the control operator
D. The control operator and the station licensee

D: When the control operator is not the station licensee, the control operator and the station licensee are both responsible for the proper operation of the station. [97.103(a)]

T1E08. Which of the following is an example of automatic control?

A. Repeater operation
B. Controlling a station over the internet
C. Using a computer or other device to send CW automatically
D. Using a computer or other device to identify automatically

A: A repeater operation may be automatically controlled. [97.3(a)(6), 97.205(d)]

T1E09. Which amateur stations may be remotely controlled?

A. Only repeater stations
B. Only automatically controlled stations
C. Only digital stations
D. Any station

D: Any amateur stations may be remotely controlled. [97.109(c)]

T1E10. Which of the following is an example of remote control as defined in Part 97?

A. A software defined radio (SDR)
B. Operating the station over the internet
C. Controlling a model aircraft, boat, or car by amateur radio
D. Earth-Moon-Earth (EME) communication

B: Operating the station over the internet is an example of remote control. [97.3(a)(39)]

T1E11. What is a control operator as defined in Part 97?

A. The person speaking or otherwise communicating messages over an amateur station
B. The person who is the licensee of an amateur station
C. An amateur operator identified in the FCC database as responsible for transmissions and FCC rules compliance at a station license location
D. An amateur operator designated by the licensee of a station to be responsible for transmissions and FCC rules compliance at that station

D: An amateur operator designated by the licensee of a station to be responsible for transmissions and FCC rules compliance at that station is the control operator as defined in Part 97. [97.3(a)(13)]

T1F – Station identification; Repeaters; Third-party communications; Club stations; FCC inspection

T1F01. When must the station licensee make the station and the station records available for inspection?

A. At any time ten days after notification by the FCC of such an inspection
B. At any time upon request by an FCC representative
C. At any time after a written notification by the FCC of such inspection
D. Only when presented with a valid warrant by an FCC official or government agent

B: The station licensee must make the station and the station records available for inspection upon request by an FCC representative. [97.103(c)]

T1F02. How often must you identify with your FCC-assigned call sign when using tactical call signs such as "Race Headquarters"?

A. Never, the tactical call is sufficient
B. Once every hour
C. At least every 10 minutes during and at the end of a communication
D. At the end of every transmission

C: When using tactical identifiers, a station must transmit its FCC-assigned call sign every 10 minutes. [97.119(a)]

T1F03. When are you required to transmit your assigned call sign?

A. At the beginning of each contact, and every 10 minutes thereafter
B. At least once during each transmission
C. At least every 15 minutes during and at the end of a communication
D. At least every 10 minutes during and at the end of a communication

D: An amateur station is required to transmit its assigned call sign at least every 10 minutes during and at the end of a communication. [97.119(a)]

T1F04. What language must you use for identification when using a phone emission?

A. Any language recognized by the United Nations
B. Any language recognized by the ITU
C. English
D. English, French, or Spanish

C: English is the acceptable language for station identification use when operating in a phone sub-band. [97.119(b)(2)]

T1F05. What method of call sign identification is required for a station transmitting phone signals?

A. Send the call sign followed by the indicator RPT
B. Send the call sign using a CW or phone emission
C. Send the call sign followed by the indicator R
D. Send the call sign using only a phone emission

B: When a station is transmitting phone signals, it should use the call sign identification system of sending the call sign using CW or phone emission. [97.119(b)(2)]

T1F06. Which of the following self-assigned indicators are acceptable when using a phone transmission?

A. KL7CC stroke W3
B. KL7CC slant W3
C. KL7CC slash W3
D. All these choices are correct

D: For identification during a phone transmission, KL7CC stroke W3, KL7CC slant W3, and KL7CC slash W3 would all be acceptable formats for a self-assigned indicator. [97.119(c)]

T1F07. Which of the following restrictions apply when a non-licensed person speaks to a foreign amateur radio station via a station under the control of an FCC-licensed amateur radio operator?

A. The person must be a U.S. citizen
B. The foreign station must be in a country with which the U.S. has a third-party agreement
C. The licensed control operator must do the station identification
D. All these choices are correct

B: An amateur station may transmit messages for a third party to any station within the jurisdiction of any foreign government when transmitting emergency or disaster relief communications and any station within the jurisdiction of any foreign government whose administration has made arrangements with the United States to allow amateur stations to be used for transmitting international communications on behalf of third parties. No station shall transmit messages for a third party to any station within the jurisdiction of any foreign government whose administration has not made such an arrangement. [97.115(a)(2)]

T1F08. What is the definition of third-party communications?

A. A message from a control operator to another amateur station control operator on behalf of another person
B. Amateur radio communications where three stations are in communications with one another
C. Operation when the transmitting equipment is licensed to a person other than the control operator
D. Temporary authorization for an unlicensed person to transmit on the amateur bands for technical experiments

A: Third-party communications are any message from the control operator (first party) of an amateur station to another amateur station control operator (second party) on behalf of another person (third party). [97.3(a)(47)]

T1F09. What type of amateur station simultaneously retransmits the signal of another amateur station on a different channel or channels?

A. Beacon station
B. Remote control station
C. Repeater station
D. Message forwarding station

C: A repeater station is an amateur station that simultaneously retransmits the signal of another amateur station on a different channel or channels. [97.3(a)(40)]

T1F10. Who is accountable if a repeater inadvertently retransmits communications that violate the FCC rules?

A. The control operator of the originating station
B. The control operator of the repeater
C. The owner of the repeater
D. Both the originating station and the repeater owner

A: If a repeater inadvertently retransmits communications that violate the FCC rules, the control operator of the originating station is accountable. [97.205(g)]

T1F11. Which of the following is a requirement for the issuance of a club station license grant?

A. The trustee must have an Amateur Extra Class operator license grant
B. The club must have at least four members
C. The club must be registered with the American Radio Relay League (ARRL)
D. All these choices are correct

B: A club station license grant may be held only by the person who is the license trustee designated by an officer of the club. The trustee must be a person who holds an operator/primary station license grant. The club must be composed of at least four persons and must have a name, a document of organization, management, and a primary purpose devoted to amateur service activities consistent with this part. [97.5(b)(2)]

T2 – Operating Procedures

T2A – Station Operation: Choosing an Operating Frequency, Calling Another Station, Test Transmissions; Band Plans: Calling Frequencies, Repeater Offsets

T2A01. What is a common repeater frequency offset in the 2-meter band?

A. Plus or minus 5 MHz
B. Plus or minus 600 kHz
C. Plus or minus 500 kHz
D. Plus or minus 1 MHz

B: The most common repeater frequency offset in the 2-meter band is plus or minus 600 kHz.

T2A02. What is the national calling frequency for FM simplex operations in the 2-meter band?

A. 146.520 MHz
B. 145.000 MHz
C. 432.100 MHz
D. 446.000 MHz

A: 146.520 and 446.000 are both simplex frequencies, but only 146.520 is in the 2m band.

T2A03. What is a common repeater frequency offset in the 70-centimeter band?

A. Plus or minus 5 MHz
B. Plus or minus 600 kHz
C. Plus or minus 500 kHz
D. Plus or minus 1 MHz

A: A common repeater frequency offset in the 70 cm band is plus or minus 5 MHz.

T2A04. What is an appropriate way to call another station on a repeater if you know the other station's call sign?

A. Say "break, break," then the other station's call sign, followed by your call sign
B. Say the station's call sign, then identify with your call sign
C. Say "CQ" three times, then the other station's call sign, followed by your call sign
D. Wait for the station to call CQ, then answer with your call sign

B: An appropriate way to call another station on a repeater if you know the other station's call sign is to say the station's call sign and then identify with your call sign.

T2A05. How should you respond to a station calling CQ?

A. Transmit "CQ" followed by the other station's call sign
B. Transmit your call sign followed by the other station's call sign
C. Transmit the other station's call sign followed by your call sign
D. Transmit a signal report followed by your call sign

C: When responding to a call of CQ, you should transmit the other station's call sign followed by your call sign.

T2A06. What is an effective way to seek a call from any phone station when not using a repeater?

A. Sign your call sign once, followed by the words "listening for a call;" if there is no answer, change the frequency and repeat
B. Say "QTC," followed by "this is" and your call sign; if there is no answer, change the frequency and repeat
C. Transmit an unmodulated carrier for approximately 10 seconds, followed by "this is" and your call sign, and pause to listen; repeat as necessary
D. Repeat "CQ" a few times, followed by "this is," and your call sign, then pause to listen; repeat as necessary

D: Repeating "CQ" a few times, followed by "this is," and your call sign, then pausing to listen is an effective way to seek a call from any phone station when not using a repeater.

T2A07. What does the term "repeater offset" mean?

A. The difference between a repeater's transmit and receive frequencies
B. The repeater has a time delay to prevent interference
C. The approximately half-second delay to allow for the "squelch tail" to be removed
D. The fee charged by the repeater owner or club to provide for maintenance

A: Repeater offset is the difference between the received and transmitted frequencies of a repeater station.

T2A08. What is the meaning of the procedural signal "CQ"?

A. A shortened form of the term "Contest QSO"
B. "Communication Quality," used to indicate transmitted audio clarity
C. Only the called station should transmit
D. Calling any station

D: The meaning of the procedural signal CQ is "calling any station."

T2A09. Which of the following is a customary way to indicate a station is listening on a repeater and looking for a contact?

A. "CQ CQ" followed by the repeater's call sign
B. The station's call sign followed by the word "listening"
C. The repeater's call sign followed by the station's call sign
D. "QSY" followed by your call sign

B: A call sign followed by the word "listening" is often used in place of CQ to indicate that you are listening on a repeater.

T2A10. What is a band plan, beyond the privileges established by the FCC?

A. A voluntary guideline for using different modes or activities within an amateur band
B. A list of operating schedules
C. A list of available net frequencies
D. A plan devised by a club to indicate frequency band usage

A: A band plan, beyond the privileges established by the FCC, is a voluntary guideline for using different modes or activities within an amateur band.

T2A11. What term describes an amateur station that is transmitting and receiving on the same frequency?

A. Full duplex
B. Diplex
C. Simplex
D. Multiplex

C: Receiving and transmitting on the same frequency is always referred to as 'simplex.'

T2B – VHF/UHF Operating Practices: FM Repeater, Simplex, Reverse Splits; Access Tones: CTCSS, DTMF; DMR Operation; Resolving Operational Problems; Q Signals

T2B01. What is the purpose of the reverse function on a VHF/UHF transceiver?

A. To reduce power output
B. To increase power output
C. To listen on a repeater's input frequency
D. To listen on a repeater's output frequency

C: A VHF/UHF transceiver's "reverse" function is used to listen on a repeater's input frequency.

T2B02. What term describes the use of a sub-audible tone transmitted along with normal voice audio to open the squelch of a receiver?

A. Carrier squelch
B. Tone burst
C. DTMF
D. CTCSS

D: CTCSS is the use of a sub-audible tone transmitted with normal voice audio to open the squelch of the receiver.

T2B03. Which of the following describes a linked repeater network?

A. A network of repeaters in which signals received by one repeater are transmitted by all the repeaters in the network
B. A single repeater with more than one receiver
C. Multiple repeaters with the same control operator
D. A system of repeaters linked by APRS

A: A linked repeater network is a network of repeaters that are all repeating the signal received by one repeater of the network.

T2B04. Which of the following could be the reason you are unable to access a repeater whose output you can hear?

A. Improper transceiver offset
B. You are using the wrong CTCSS tone
C. You are using the wrong DCS code
D. All these choices are correct

D: There are several common problems that might cause you to be able to hear but not access a repeater even when transmitting with the proper offset. This may occur when the repeater receiver requires audio tone burst, a CTCSS tone, or a DCS tone sequence for access.

T2B05. Which of the following would cause your FM transmission audio to drop out on voice peaks?

A. Your repeater offset is inverted
B. Your FM deviation is too low
C. You are talking too loudly
D. Your transmit power is too high

C: Speaking too loudly is the most common cause of audio distortions on voice peaks.

T2B06. What type of signaling to a repeater uses two simultaneous audio tones?

A. DTMF
B. CTCSS
C. GMRS
D. D-STAR

A: DTMF signals use pairs of audio tones.

T2B07. How can you join a digital repeater's "talkgroup"?

A. Register with the local Frequency Coordinator
B. Register with the digital repeater control operator
C. Program your radio with the group's ID or code
D. Program your radio with the DTMF tone used by the talkgroup

C: Once you have identified the talkgroup you wish to join, all you need is the group's ID or code in order to receive and transmit on their channel.

T2B08. Which of the following applies when two stations transmitting on the same frequency interfere with each other?

A. The stations should negotiate continued use of the frequency
B. Both stations should choose another frequency to avoid conflict
C. Whichever station was on the frequency first has preemptive rights to the frequency
D. Use subaudible tones so both stations can share the frequency

A: If two stations transmitting on the same frequency begin to interfere with one another, it is up to the participants to resolve the issue. There is no absolute right to any amateur frequency, so operators are encouraged to remain courteous and work out a solution.

T2B09. Why are simplex channels designated in the VHF/UHF band plans?

A. So stations within range of each other can communicate without tying up a repeater
B. For contest operation
C. For working DX only
D. So stations with simple transmitters can access the repeater without automated offset

A: The main purpose of designating simplex channels is to allow communication directly between sender and receiver and not tie up a repeater.

T2B10. Which Q signal indicates that you are receiving interference from other stations?

A. QRM
B. QRN
C. QTH
D. QSB

A: The Q signal used to indicate that you are receiving interference from other stations is QRM. Q signals are useful methods for accelerating communication. A Q signal may be issued as a question or as a response. For example, QRG may mean "What is my exact frequency?" or "Your exact frequency is."

T2B11. Which Q signal indicates that you are changing frequency?

A. QRU
B. QSY
C. QSL
D. QRZ

B: The Q signal used to indicate that you're changing frequency is QSY. This Q signal can also be used to request a shift to another frequency.

T2B12. What is the digital color code used on DMR repeater systems?

A. An access code which must be programmed into a DMR transmitter to access a specific repeater
B. A code which automatically programs the repeater offset to a frequency chosen by the user
C. A code which identifies the specific CODEC being used so that the repeater can properly decode the audio
D. A code transmitted by the repeater to indicate health and status of the equipment

A: The digital color code on DMR repeater systems is an access code that must be programmed into a DMR transmitter to access a specific repeater.

T2B13. What is the purpose of a squelch function?

A. Reduce a CW transmitter's key clicks
B. Mute the receiver audio when a signal is not present
C. Eliminate parasitic oscillations in an RF amplifier
D. Reduce interference from impulse noise

B: A squelch function is intended to mute the receiver audio when a signal is not present.

T2B14. Which of the following is a "talkgroup"?

A. A group of amateur radio operators who share email on a specific amateur radio topic
B. An identifier used by DMR to organize radio traffic so that those who want to hear the group aren't bothered by other radio traffic
C. The members of a net organized for amateur radio operators with a specific common interest
D. A method of automatically alerting all group members to the frequency where any member is transmitting

B: A talkgroup is an identifier used by DMR to organize radio traffic so that those who want to hear the group aren't bothered by other radio traffic.

T2C – Public Service: Emergency Operations, Applicability of FCC Rules, RACES and ARES, Net and Traffic Procedures, Operating Restrictions During Emergencies, Use of Phonetics in Message Handling

T2C01. When do the FCC Part 97 Amateur Radio Service rules NOT apply to the operation of an amateur station?

A. When operating under RACES rules
B. When operating under FEMA rules
C. When operating under ARES rules
D. FCC rules always apply

D: FCC rules always apply to the operation of an amateur station. [97.103(a)]

T2C02. Which of the following are typical duties of a Net Control Station?

A. Choose the regular net meeting time and frequency
B. Ensure that all stations checking into the net are properly licensed for operation on the net frequency
C. Call the net to order and direct communications between stations checking in
D. All these choices are correct

C: A typical duty of a Net Control Station is to call the net to order and direct communications between stations checking in. The Net Control Station helps maintain order and manages the flow of traffic on the net.

T2C03. What technique is used to ensure that voice messages containing unusual words are received correctly?

A. Send the words by voice and Morse code
B. Use the Q-code "QSR" to ask the receiving station to repeat the words back
C. Spell the words using a standard phonetic alphabet
D. All these choices are correct

C: To ensure that any proper names and unusual words are copied correctly by the receiving station, it is good practice to spell them out using a standard phonetic alphabet.

T2C04. What is RACES?

A. An emergency organization combining amateur radio and citizens band operators and frequencies
B. An international radio experimentation society
C. A radio contest held in a short period, sometimes called a "sprint"
D. An FCC Part 97 amateur radio service for civil defense communications during national emergencies

D: RACES is the Radio Amateur Civil Emergency Service, an FCC Part 97 amateur radio service for civil defense communications during national emergencies. It is a specific emergency communications service defined by FCC rules.

T2C05. What does the term "traffic" refer to in net operation?

A. Formal messages exchanged by net stations
B. The number of stations checking in and out of a net
C. Operation by mobile or portable stations
D. A count of the number of activations of the net each month

A: Traffic is a special term for messages on a network.

T2C06. What is the Amateur Radio Emergency Service (ARES)?

A. A group of licensed amateurs who have voluntarily registered their qualifications and equipment for communications duty in the public service
B. A group of licensed amateurs who are members of the military and who voluntarily agreed to provide message handling services in the case of an emergency
C. A training program that provides licensing courses for those interested in obtaining an amateur license to use during emergencies
D. A training program that certifies amateur operators for membership in the Radio Amateur Civil Emergency Service

A: Amateur Radio Emergency Service (ARES) are licensed amateurs who have voluntarily registered their qualifications and equipment for communications duty in the public service

T2C07. Which of the following is standard practice when you participate in a net?

A. When first responding to the net control station, transmit your call sign, name, and address as in the FCC database
B. Record the time of each of your transmissions
C. Unless you are reporting an emergency, transmit only when directed by the net control station
D. All these choices are correct

C: To minimize disruptions to an emergency traffic net once you have checked in, do not transmit on the net frequency until asked to do so by the net control station.

T2C08. Which of the following relays messages using email addresses based on amateur callsigns?

A. Winlink
B. FT8
C. PSK31
D. AMTOR

A: Winlink relays messages using email addresses based on amateur callsigns.

T2C09. Are amateur station control operators ever permitted to operate outside the frequency privileges of their license class?

A. No
B. Yes, but only when part of a FEMA emergency plan
C. Yes, but only when part of a RACES emergency plan
D. Yes, but only in situations involving the immediate safety of human life or protection of property

D: Amateur station control operators are permitted to operate outside the frequency privileges of their license class only if necessary in situations involving the immediate safety of human life or protection of property

T2C10. What information is contained in the preamble of a formal traffic message?

A. The email address of the originating station
B. The address of the intended recipient
C. The telephone number of the addressee
D. Information needed to track the message

D: The preamble in a formal traffic message is information needed to track the message as it passes through the amateur radio traffic handling system. The preamble is the first part of a radiogram. It identifies the source and some basic information about the message.

T2C11. What is meant by "check" in a radiogram header?

A. The number of words or word equivalents in the text portion of the message
B. The call sign of the originating station
C. A list of stations that have relayed the message
D. A box on the message form that indicates that the message was received and/or relayed

A: The term "check," in reference to a radiogram header, is a count of the number of words or word equivalents in the text portion of the message. It is important to confirm the word count with the receiving station, as an incorrect word count will suggest to relaying stations that there is an error in the message.

T2C12. Which of the following requires certification by a civil defense agency?

A. ARES
B. RACES
C. MARS
D. SKYWARN

B: RACES requires certification by a civil defense agency.

T3 – Radio Wave Propagation

T3A – Radio wave characteristics: how a radio signal travels, fading, multipath, polarization, wavelength vs absorption; Antenna orientation

T3A01. Why do VHF signal strengths sometimes vary greatly when the antenna is moved only a few feet?

A. The signal path encounters different concentrations of water vapor
B. VHF ionospheric propagation is very sensitive to path length
C. Multipath propagation cancels or reinforces signals
D. The Doppler effect causes slight frequency shifts which result in changes in signal strength

C: If a signal is experiencing multipath propagation, it can easily cancel itself out or reinforce itself over a distance of a few feet.

T3A02. How does vegetation affect UHF and microwave signals?

A. Causes knife-edge diffraction, distorting voice peaks
B. Absorbs signals, leading to poor reception of weak signals
C. Amplifies signals, improving reception of weak signals
D. Has no effect

B: Vegetation can absorb UHF and microwave signals, significantly weakening them in areas with dense vegetation.

T3A03. What antenna polarization is normally used for long-distance CW and SSB contacts on the VHF and UHF bands?

A. Right-hand circular
B. Left-hand circular
C. Horizontal
D. Vertical

C: A horizontal antenna polarization is normally used for long-distance weak-signal CW and SSB contacts using the VHF and UHF bands.

T3A04. What is the effect of antenna cross-polarization over a line-of-sight VHF or UHF path?

A. Modulation sidebands might become inverted
B. Received signal strength is reduced
C. Signals have an echo effect
D. Nothing significant will happen

B: If the antennas at opposite ends of a VHF or UHF line of sight radio link are not using the same polarization, the signals could be significantly weaker.

T3A05. When using a directional antenna, how might your station be able to communicate with a distant repeater if buildings or obstructions are blocking the direct line of sight path?

A. Change from vertical to horizontal polarization
B. Try to find a path that reflects signals to the repeater
C. Try the long path
D. Increase the antenna SWR

B: When using a directional antenna, a station might be able to access a distant repeater if buildings or obstructions are blocking the direct line of sight path by trying to find a path that reflects signals to the repeater.

T3A06. What is the meaning of the term "picket fencing"?

A. Alternating transmissions during a net operation
B. Rapid flutter on mobile signals due to multipath propagation
C. A type of ground system used with vertical antennas
D. Interference from cable TV in the form of carriers at fixed intervals across the band

B: Picket fencing is a term used to describe the rapid fluttering sound sometimes heard from mobile stations that are moving while transmitting.

T3A07. What weather condition might decrease range at microwave frequencies?

A. High winds
B. Low barometric pressure
C. Precipitation
D. Colder temperatures

C: Precipitation can decrease the effective range of microwave frequencies.

T3A08. What is a likely cause of irregular fading of signals propagated by the ionosphere?

A. Frequency shift due to Faraday rotation
B. Interference from thunderstorms
C. Intermodulation distortion
D. Random combining of signals arriving via different paths

D: The cause of irregular fading of signals from distant stations during times of generally good reception is the random combining of signals arriving via different path lengths.

T3A09. Which of the following results from the fact that signals propagated by the ionosphere are elliptically polarized?

A. Digital modes are unusable
B. Either vertically or horizontally polarized antennas may be used for transmission or reception
C. FM voice is unusable
D. Both the transmitting and receiving antennas must be of the same polarization

B: Because signals propagated by the ionosphere are elliptically polarized, either vertically or horizontally polarized antennas can be used to transmit and receive them.

T3A10. What effect does multi-path propagation have on data transmissions?

A. Transmission rates must be increased by a factor equal to the number of separate paths observed
B. Transmission rates must be decreased by a factor equal to the number of separate paths observed
C. No significant changes will occur if the signals are transmitted using FM
D. Error rates are likely to increase

D: If the data signals propagate over multiple paths, error rates are likely to increase.

T3A11. Which region of the atmosphere can reflect HF radio waves?

A. The stratosphere
B. The troposphere
C. The ionosphere
D. The electrosphere

C: The ionosphere is the region of the atmosphere that can reflect HF radio waves.

T3A12. What effect does fog or rain have on 10-meter and 6-meter band signals?

A. Absorption
B. Little effect
C. Deflection
D. Increased range

B: Fog or rain has little effect on 10-meter and 6-meter band signals.

T3B – Electromagnetic wave properties: wavelength vs frequency, nature and velocity of electromagnetic waves, relationship of wavelength and frequency; Electromagnetic spectrum definitions: UHF, VHF, HF

T3B01. What is the relationship between the electric and magnetic fields of an electromagnetic wave?

A. They travel at different speeds
B. They are in parallel
C. They revolve in opposite directions
D. They are at right angles

D: The electric and magnetic fields of electromagnetic waves are at right angles (perpendicular) to each other.

T3B02. What property of a radio wave defines its polarization?

A. The orientation of the electric field
B. The orientation of the magnetic field
C. The ratio of the energy in the magnetic field to the energy in the electric field
D. The ratio of the velocity to the wavelength

A: The orientation of a radio wave's electric field determines its polarization.

T3B03. What are the two components of a radio wave?

A. Impedance and reactance
B. Voltage and current
C. Electric and magnetic fields
D. Ionizing and non-ionizing radiation

C: The two components of a radio wave are electric and magnetic fields. This is because a radio wave is an electromagnetic wave.

T3B04. What is the velocity of a radio wave traveling through free space?

A. Speed of light
B. Speed of sound
C. 0.86 times the speed of light
D. 1.86 times the speed of sound

A: A radio wave travels through free space at the speed of light. This is approximately 300,000,000 meters per second.

T3B05. What is the relationship between wavelength and frequency?

A. Wavelength gets longer as frequency increases
B. Wavelength gets shorter as frequency increases
C. Wavelength is constant at all frequencies
D. Wavelength and frequency increase as path length increases

B: The wavelength of a radio wave gets shorter as the frequency increases. That is, these two characteristics are inversely proportional. If the wavelength grows longer, the frequency will decrease.

T3B06. What is the formula for converting frequency to approximate wavelength in meters?

A. Wavelength in meters equals frequency in hertz multiplied by 300
B. Wavelength in meters equals frequency in hertz divided by 300
C. Wavelength in meters equals frequency in megahertz divided by 300
D. Wavelength in meters equals 300 divided by frequency in megahertz

D: The formula for converting frequency to wavelength in meters is wavelength equals 300 divided by frequency in megahertz.

T3B07. In addition to frequency, which of the following is used to identify amateur radio bands?

A. The approximate wavelength in meters
B. Traditional letter/number designators
C. Channel numbers
D. All these choices are correct

A: The approximate wavelength (in meters) of radio waves is often used to identify the different frequency bands.

T3B08. What frequency range is referred to as VHF?

A. 30 kHz to 300 kHz
B. 30 MHz to 300 MHz
C. 300 kHz to 3000 kHz
D. 300 MHz to 3000 MHz

B: The frequency limits of the VHF spectrum are 30 to 300 MHz.

T3B09. What frequency range is referred to as UHF?

A. 30 to 300 kHz
B. 30 to 300 MHz
C. 300 to 3000 kHz
D. 300 to 3000 MHz

D: The UHF spectrum is from 300 to 3000 MHz.

T3B10. What frequency range is referred to as HF?

A. 300 to 3000 MHz
B. 30 to 300 MHz
C. 3 to 30 MHz
D. 300 to 3000 kHz

C: The frequency range of 3 to 30 MHz is referred to as HF.

T3B11. What is the approximate velocity of a radio wave in free space?

A. 150,000,000 meters per second
B. 300,000,000 meters per second
C. 300,000,000 miles per hour
D. 150,000,000 miles per hour

B: The approximate velocity of a radio wave as it travels through free space is 300,000,000 m/s.

T3B12. Which of these frequencies travels at the highest velocity in free space?

A. Microwaves
B. UHF
C. VHF
D. All radio frequencies travel at the same velocity

D: All radio frequencies travel at the same velocity in free space.

T3C – Propagation modes: sporadic E, meteor scatter, auroral propagation, tropospheric ducting; F region skip; Line of sight and radio horizon

T3C01. Why are simplex UHF signals rarely heard beyond their radio horizon?

A. They are too weak to go very far
B. FCC regulations prohibit them from going more than 50 miles
C. UHF signals are usually not propagated by the ionosphere
D. UHF signals are absorbed by the ionospheric D region

C: Simplex (that is, not by means of a repeater) UHF signals are rarely heard from stations outside the local coverage area because UHF signals are usually not reflected by the ionosphere.

T3C02. What is a characteristic of HF communication compared with communications on VHF and higher frequencies?

A. HF antennas are generally smaller
B. HF accommodates wider bandwidth signals
C. Long-distance ionospheric propagation is far more common on HF
D. There is less atmospheric interference (static) on HF

C: The main benefit to HF radio waves is that they typically bounce off the ionosphere and therefore can have a much greater range. Higher frequencies penetrate the ionosphere and therefore aren't propagated.

T3C03. What is one characteristic of VHF signals received via auroral backscatter?

A. They are often received from 10,000 miles or more
B. They are distorted, with a characteristic raspy sound
C. They occur only during winter nighttime hours
D. They are generally strongest when your antenna is aimed west

B: A characteristic of VHF signals received via auroral backscatter is that the signals exhibit rapid fluctuations of strength and often sound distorted.

T3C04. Which of the following types of propagation is most commonly associated with occasional strong signals on the 10-, 6-, and 2-meter bands from beyond the radio horizon?

A. Backscatter
B. Sporadic E
C. D region absorption
D. Gray-line propagation

B: The sporadic E propagation type is most commonly associated with occasional strong signals on the 10, 6, and 2 meter bands.

T3C05. Which of the following effects may allow radio signals to travel beyond obstructions between the transmitting and receiving stations?

A. Knife-edge diffraction
B. Faraday rotation
C. Quantum tunneling
D. Doppler shift

A: A knife-edge diffraction might allow radio signals to be heard despite obstructions between the transmitting and receiving systems.

T3C06. What type of propagation is responsible for allowing over-the-horizon VHF and UHF communications to ranges of approximately 300 miles on a regular basis?

A. Tropospheric ducting
B. D region refraction
C. F2 region refraction
D. Faraday rotation

A: Tropospheric ducting is the type of propagation that regularly allows VHF and UHF signals to travel well beyond the horizon for distances of about 300 miles. It is caused by atmospheric conditions, often involving temperature inversions.

T3C07. What band is best suited for communicating via meteor scatter?

A. 33 centimeters
B. 6 meters
C. 2 meters
D. 70 centimeters

B: The 6-meter band is best suited to communicating via meteor scatter.

T3C08. What causes tropospheric ducting?

A. Discharges of lightning during electrical storms
B. Sunspots and solar flares
C. Updrafts from hurricanes and tornadoes
D. Temperature inversions in the atmosphere

D: Tropospheric ducting is caused by temperature inversions in the atmosphere. This is a type of very high frequency propagation that occurs when warm air passes over cold air.

T3C09. What is generally the best time for long-distance 10-meter band propagation via the F region?

A. From dawn to shortly after sunset during periods of high sunspot activity
B. From shortly after sunset to dawn during periods of high sunspot activity
C. From dawn to shortly after sunset during periods of low sunspot activity
D. From shortly after sunset to dawn during periods of low sunspot activity

A: Long-distance 10-meter band propagation via the F region is best conducted between dawn and shortly after sunset during periods of high sunspot activity.

T3C10. Which of the following bands may provide long-distance communications via the ionosphere's F region during the peak of the sunspot cycle?

A. 6 and 10 meters
B. 23 centimeters
C. 70 centimeters and 1.25 meters
D. All these choices are correct

A: During periods of high sunspot activity, the 6 meter and 10 meter bands may be used for long-distance communication.

T3C11. Why is the radio horizon for VHF and UHF signals more distant than the visual horizon?

A. Radio signals move somewhat faster than the speed of light
B. Radio waves are not blocked by dust particles
C. The atmosphere refracts radio waves slightly
D. Radio waves are blocked by dust particles

C: VHF and UHF radio signals usually travel somewhat farther than the visual line of sight distance between two stations because the atmosphere refracts radio waves slightly.

T4 – Amateur Radio Practices

T4A – Station setup: connecting a microphone, RF power meter, a power source, a computer, digital equipment, an SWR meter; RF grounding; Mobile radio installation

T4A01. Which of the following is an appropriate power supply rating for a typical 50-watt output mobile FM transceiver?

A. 24.0 volts at 4 amperes
B. 13.8 volts at 4 amperes
C. 24.0 volts at 12 amperes
D. 13.8 volts at 12 amperes

D: For a typical 50 watt output mobile FM transceiver, 13.8 volts at 12 amperes would be an appropriate power supply rating.

T4A02. Which of the following should be considered when selecting an accessory SWR meter?

A. The frequency and power level at which the measurements will be made
B. The distance that the meter will be located from the antenna
C. The maximum SWR anticipated on the transmission line
D. The ability of the meter to compensate for a poor transmission line match to the antenna

A: The frequency and power level at which the measurements will be made should be considered when selecting an accessory SWR meter.

T4A03. Why are short, heavy-gauge wires used for a transceiver's DC power connection?

A. To minimize voltage drop when transmitting
B. To provide a close match to the power supply output impedance
C. To avoid RF interference
D. To minimize radiative losses in the power cable

A: Heavier gauge wire and short lengths will reduce the resistance as much as possible, thus maintaining sufficient voltage for peak performance.

T4A04. How are the audio input and output of a transceiver connected in a station configured to operate using FT8?

A. To a computer running a terminal program and connected to a terminal node controller unit
B. To the audio output and input of a computer running FT8 software
C. To an FT8 conversion unit, a keyboard, and a computer monitor
D. To a computer connected to the FT8converter.com website

B: In a station configured to operate using FT8, the transceiver audio input and output are connected to the audio input and output of a computer running WSJT-X software.

T4A05. Where should an RF power meter be installed?

A. In the feed line, between the transmitter and antenna
B. At the power supply output
C. In parallel with the push-to-talk line and the antenna
D. In the power supply cable, as close as possible to the radio

A: An RF power meter should be installed in the feed line, between the transmitter and antenna.

T4A06. What signals are used in a computer-radio interface for digital mode operation?

A. Receive and transmit mode, status, and location
B. Antenna and RF power
C. Receive audio, transmit audio, and transmitter keying
D. NMEA GPS location and DC power

C: The connections needed are sending and receiving audio in conjunction with transmitter keying in order to allow the computer to trigger the transmitter automatically.

T4A07. Which of the following is one of the connections required between a computer and a transceiver to operate digital modes?

A. Computer "line out" to transceiver push-to-talk
B. Computer "line in" to transceiver push-to-talk
C. Computer "line in" to transceiver speaker connector
D. Computer "line out" to transceiver speaker connector

C: Connecting the computer "line in" to transceiver speaker connector enables operation of digital modes.

T4A08. Which of the following conductors is preferred for bonding at RF?

A. Copper braid removed from coaxial cable
B. Copper-clad steel wire
C. Twisted-pair cable
D. Flat copper strap

D: A flat strap conductor is the best to use for RF grounding. Most ham radio systems have a special terminal for grounding. Copper strap is most often used for grounding.

T4A09. How can you determine the length of time that equipment can be powered from a battery?

A. Divide the watt-hour rating of the battery by the peak power consumption of the equipment
B. Divide the battery ampere-hour rating by the average current draw of the equipment
C. Multiply the watts per hour consumed by the equipment by the battery power rating
D. Multiply the square of the current rating of the battery by the input resistance of the equipment

B: To determine how long equipment can be powered from a battery, divide the battery's ampere-hour rating by the average current draw of the equipment. This gives the approximate operating time in hours.

T4A10. What function does a digital mode hotspot perform for nearby transceivers?

A. Communication with a digital voice or data network
B. FT8 digital communications via AFSK using a smartphone connected to the internet
C. RTTY encoding and decoding without a computer
D. High-speed digital communications for meteor scatter

A: Communication using digital voice or data systems via the internet is performed with a transceiver and a digital mode hot spot.

T4A11. Where should the negative power return of a mobile transceiver be connected in a vehicle?

A. At the 12-volt battery chassis ground
B. At the antenna mount
C. To any metal part of the vehicle
D. Through the transceiver's mounting bracket

A: A mobile transceiver's power negative connection should be made at the 12 volt battery chassis ground.

T4A12. What is an electronic keyer?

A. A device for switching antennas from transmit to receive
B. A device for voice activated switching from receive to transmit
C. A device that assists in manual sending of Morse code
D. An interlock to prevent unauthorized use of a radio

C: An electronic keyer automatically generates the dits and dahs of Morse code. This is a key component of some CW/data setups.

T4B – Operating controls: frequency tuning, use of filters, squelch function, AGC, memory channels, noise blanker, microphone gain, receiver incremental tuning (RIT), bandwidth selection, scanning function; Digital transceiver configuration; DMR code plugs and talk groups

T4B01. What is the effect of excessive microphone gain on SSB transmissions?

A. Frequency instability
B. Distorted transmitted audio
C. Increased SWR
D. Sideband inversion

B: If a transmitter is operated with the microphone gain set too high, the output signal might become distorted.

T4B02. Which of the following can be used to enter a transceiver's operating frequency?

A. The keypad or VFO knob
B. The CTCSS or DTMF encoder
C. The Automatic Frequency Control
D. All these choices are correct

A: The keypad or the VFO knob can be used to enter the operating frequency on a modern transceiver. A transceiver is a single unit that contains both a radio transmitter and receiver.

T4B03. How is squelch adjusted so that a weak FM signal can be heard?

A. Set the squelch threshold so that receiver output audio is on all the time
B. Turn up the audio level until it overcomes the squelch threshold
C. Turn on the anti-squelch function
D. Enable squelch enhancement

A: Setting the squelch threshold so that receiver output audio is on all the time will help a weak FM signal be heard.

T4B04. What does an FM signal sound like when received slightly off frequency?

A. The audio increases in pitch
B. The audio decrease in pitch
C. There is no effect except for reduction in amplitude
D. The audio becomes distorted

D: An FM signal sounds distorted when it is received slightly off frequency.

T4B05. What does the scanning function of an FM transceiver do?

A. Checks incoming signal deviation
B. Prevents interference to nearby repeaters
C. Tunes through a range of frequencies to check for activity
D. Tunes through a range of frequencies to determine the antenna's resonant frequency

C: The scanning function of an FM transceiver tunes through a range of frequencies to check for activity.

T4B06. Which of the following controls could be used if the voice pitch of a single-sideband signal returning to your CQ call seems too high or low?

A. The AGC or limiter
B. The bandwidth selection
C. The tone squelch
D. The RIT or Clarifier

D: The RIT or clarifier controls could be used if the voice pitch of a single-sideband signal seems too high or low. Receiver incremental tuning allows the receiver to be tuned up to 3 kilohertz on either side of the transmitter frequency.

T4B07. What is a DMR "code plug"?

A. An adapter cable used to connect a DMR radio to a computer for internet access
B. Configuration data loaded onto your radio to access repeaters and talkgroups
C. An upgrade to DMR programming software provided by the radio manufacturer to accommodate new radio models
D. A Coder-Decoder (CODEC) that converts analog voice data to DMR digital data and vice versa

B: A DMR "code plug" contains configuration data loaded onto your radio to access repeaters and talkgroups.

T4B08. What is the advantage of having a choice of receiver filter bandwidths in a multimode transceiver?

A. Permits monitoring several modes simultaneously by selecting a separate filter for each mode
B. Permits noise or interference reduction by selecting a bandwidth matching the mode
C. Increases the number of frequencies that can be stored in memory
D. Increases the amount of offset between receive and transmit frequencies

B: The advantage of having a choice of receiver filter bandwidths in a multimode transceiver is that it permits noise or interference reduction by selecting a bandwidth that matches the mode being used. Narrower or wider filters can improve reception depending on the signal type.

T4B09. How is a specific group of stations selected on a DMR digital voice transceiver?

A. By retrieving the frequencies from transceiver memory
B. By enabling the group's CTCSS tone
C. By entering the group's identification code
D. By inserting a five-pin, pre-programmed code plug

C: A specific group of stations is selected on a digital voice transceiver by entering the group's identification code.

T4B10. Which of the following receiver filter bandwidths provides the best signal-to-noise ratio for SSB reception?

A. 500 Hz
B. 1000 Hz
C. 2400 Hz
D. 5000 Hz

C: 2400 Hz is an appropriate receive filter to select in order to provide the best signal-to-noise ratio for SSB reception.

T4B11. Which of the following must be programmed into a D-STAR digital transceiver before transmitting?

A. Your call sign
B. Your output power
C. The codec type being used
D. All these choices are correct

A: Your call sign must be programmed into a D-STAR digital transceiver before transmitting.

T5 – Electrical Principles

T5A – Current and Voltage: Terminology and Units, Conductors and Insulators, Alternating and Direct Current

T5A01. Electrical current is measured in which of the following units?

A. Volts
B. Watts
C. Ohms
D. Amperes

D: Electrical current is measured in amperes, usually referred to as amps. Current is the amount of electrons flowing through a circuit.

T5A02. Electrical power is measured in which of the following units?

A. Volts
B. Watts
C. Watt-hours
D. Amperes

B: Electrical power is measured in watts. Electrical power is the rate at which electricity can be converted into other types of energy. It is calculated by multiplying the current and the voltage drop. A watt is equal to one joule per second.

T5A03. What is the term for the flow of electrons in an electric circuit?

A. Voltage
B. Amperes per second
C. Volts per second
D. Current

D: The flow of electrons in an electric circuit is the current. Current is measured in amperes, also known as amps.

T5A04. What term describes the number of times per second that an alternating current makes a complete cycle?

A. Pulse rate
B. Wave number
C. Frequency
D. Wavelength

C: Frequency describes the number of times per second that an alternating current makes a complete cycle.

T5A05. A difference in which of the following causes electron flow?

A. Voltage
B. Ampere-hours
C. Capacitance
D. Inductance

A: The electrical term for the electromotive force (EMF) that causes electron flow is voltage. It is useful to think of voltage as the electrical equivalent of water pressure.

T5A06. What is the unit of frequency?

A. Hertz
B. Henry
C. Farad
D. Epicycles per second

A: Hertz is the unit of frequency.

T5A07. Why are metals generally good conductors of electricity?

A. They have relatively high density
B. They have many free electrons
C. They have many free protons
D. They have relatively high Young's modulus values

B: Metals generally good conductors of electricity because they have many free electrons. This phenomenon of the free electrons is also called the "sea of electrons."

T5A08. Which of the following is a good electrical insulator?

A. Sea water
B. Glass
C. Stainless steel
D. Graphite

B: Glass is a good electrical insulator.

T5A09. Which of the following describes alternating current?

A. Current that alternates between a positive direction and zero
B. Current that alternates between a negative direction and zero
C. Current that alternates between positive and negative directions
D. All these answers are correct

C: As the name implies, alternating current (AC) is current that alternates between positive and negative directions.

T5A10. Which term describes the rate at which electrical energy is used?

A. Resistance
B. Current
C. Power
D. Voltage

C: The rate at which electrical energy is used is power.

T5A11. What type of current flow is opposed by resistance?

A. Direct current
B. Alternating current
C. RF current
D. All these choices are correct

D: All current flow is opposed by resistance.

T5B – Math for Electronics: Conversion of Electrical Units, Decibels

T5B01. How many milliamperes is 1.5 amperes?

A. 0.0000015 milliamperes
B. 0.0015 milliamperes
C. 1500 milliamperes
D. 1,500,000 milliamperes

C: 1.5 amperes is equivalent to 1500 mA. The prefix milli- means thousandth, so a milliampere is a thousandth of an ampere. Therefore, 1.5 amperes can be converted to milliamperes by multiplying by a thousand: 1.5 x 1000 = 1500.

T5B02. Which is equal to 1,500,000 hertz?

A. 1500 kHz
B. 1500 MHz
C. 15 GHz
D. 150 kHz

A: Another way to specify a radio signal frequency of 1,500,000 Hz is 1500 kHz. In the metric system, the prefix kilo- means thousand, so a kilohertz is equal to 1000 hertz. Therefore, 1,500,000 Hz can be converted to kHz by dividing by a thousand: 1,500,000/1000 = 1500 kHz.

T5B03. Which is equal to one kilovolt?

A. One one-thousandth of a volt
B. One hundred volts
C. One thousand volts
D. One million volts

C: 1000 V is equal to 1 kV. The metric prefix kilo- means thousand, so a thousand volts is the same as one kilovolt. Therefore, V can be converted to kV by dividing by a thousand: 1000/1000 = 1.

T5B04. Which is equal to one microvolt?

A. One one-millionth of a volt
B. One million volts
C. One thousand kilovolts
D. One one-thousandth of a volt

A: 1/1,000,000 of a volt is equal to 1 μV. The metric prefix micro- means millionth, so 1,000,000 microvolts is equal to one volt. Volts can be converted to microvolts by dividing by a million: 1/1,000,000 = 1 μV.

T5B05. Which is equal to 500 milliwatts?

A. 5 watts
B. 0.5 watts
C. 500,000 watts
D. 500,000,000 watts

B: 0.5 W is equivalent to 500 mW. The metric prefix milli- means thousandth, so 1000 mW is the same as 1 W. This conversion can be accomplished by multiplying by a thousand: 0.5 x 1000 = 500.

T5B06. Which is equal to 3000 milliamperes?

A. 0.003 amperes
B. 0.3 amperes
C. 3,000,000 amperes
D. 3 amperes

D: A 3000-milliampere current is equivalent to 3 amperes. The metric prefix milli- means thousandth, so milliamperes can be converted to amperes by dividing by a thousand. An ammeter is an instrument used to measure electric current.

T5B07. Which is equal to 3.525 MHz?

A. 0.003525 kHz
B. 35.25 kHz
C. 3525 kHz
D. 3,525,000 kHz

C: A frequency of 3.525 MHz is equivalent to 3525 kilohertz. The metric prefix mega- means millions, and the prefix kilo- means thousands. Millions are the same as 10^6, while thousands are equal to 10^3. Megahertz can be converted into kilohertz, then, by multiplying by 10^3, or a thousand: 3.525 MHz x 1000 = 3525.

T5B08. Which is equal to 1,000,000 picofarads?

A. 0.001 microfarads
B. 1 microfarad
C. 1000 microfarads
D. 1,000,000,000 microfarads

B: 1,000,000 picofarads is equivalent to 1 microfarad. The metric prefix pico- means trillionths (10^{-12}), while the prefix micro-means millionths (10^{-6}). Picofarads can be converted to microfarads, then, by dividing by 10^{6}, or one million: 1,000,000/1,000,000 = 1.

T5B09. Which decibel value most closely represents a power increase from 5 watts to 10 watts?

A. 2 dB
B. 3 dB
C. 5 dB
D. 10 dB

B: A power increase from 5 to 10 watts creates approximately 3 dB of change. The relationship between power increases and changes in decibel levels is complex. Whenever the power is doubled, however, the decibel level will rise by three. If the power is quadrupled, the decibel level rises by 6. If the power increases by ten times, the decibel level also rises by ten.

T5B10. Which decibel value most closely represents a power decrease from 12 watts to 3 watts?

A. -1 dB
B. -3 dB
C. -6 dB
D. -9 dB

C: A power decrease from 12 watts to 3 watts would result in approximately 6 dB of change. Just as a quadrupling of power would result in an extra 6 dB, so will a quartering of power cause the decibel level to drop by 6.

T5B11. Which decibel value represents a power increase from 20 watts to 200 watts?

A. 10 dB
B. 12 dB
C. 18 dB
D. 28 dB

A: A power increase from 20 W to 200 W would result in approximately 10 dB of change. When the power increases by ten times, the result is 10 dB of change.

T5B12. Which is equal to 28400 kHz?

A. 28.400 kHz
B. 2.800 MHz
C. 284.00 MHz
D. 28.400 MHz

D: One megahertz (MHz) is equivalent to one thousand kilohertz (kHz), so to convert a frequency from kilohertz to megahertz, divide the kilohertz number by one thousand.

T5B13. Which is equal to 2425 MHz?

A. 0.002425 GHz
B. 24.25 GHz
C. 2.425 GHz
D. 242.5 GHz

C: 2425 MHz is equal to 2.425 GHz.

T5C – Capacitance and Inductance Terminology and Units; Radio Frequency Definition and Units; Impedance Definition and Units; Calculating Power

T5C01. What describes the ability to store energy in an electric field?

A. Inductance
B. Resistance
C. Frequency
D. Capacitance

D: The ability to store energy in an electric field is called capacitance. Capacitance can also be defined as the ratio of the electric charge that is shifted from one conductor to another to the resulting potential difference between them.

T5C02. What is the unit of capacitance?

A. Farad
B. Ohm
C. Volt
D. Henry

A: The basic unit of capacitance is the farad. One farad is equal to the capacitance of a capacitor that has an equal and opposite charge of 1 coulomb on each of its plates, and 1 V of potential difference between those plates.

T5C03. What describes the ability to store energy in a magnetic field?

A. Admittance
B. Capacitance
C. Resistance
D. Inductance

D: The ability to store energy in a magnetic field is called inductance. Inductance can also be defined as the characteristic of an electric circuit that enables electromagnetic force to be created in it through a varying magnetic flux.

T5C04. What is the unit of inductance?

A. Coulomb
B. Farad
C. Henry
D. Ohm

C: The basic unit of inductance is the henry.

T5C05. What is the unit of impedance?

A. Volt
B. Ampere
C. Coulomb
D. Ohm

D: The ohm is the unit of impedance.

T5C06. What is the abbreviation for kilohertz?

A. kHz
B. khz
C. KHz
D. KHZ

A: kHz is the abbreviation for kilohertz.

T5C07. What is the abbreviation for megahertz?

A. mHz
B. mHZ
C. Mhz
D. MHz

D: MHz is the abbreviation for megahertz.

T5C08. What formula is used to calculate electrical power (P) in a DC circuit?

A. P = I × E
B. P = E / I
C. P = I2 × E
D. P = I / E

A: The formula used to calculate electrical power in a DC circuit is power equals voltage multiplied by current.

T5C09. How much power is delivered by a voltage of 13.8 volts DC and a current of 10 amperes?

A. 138 watts
B. 0.7 watts
C. 23.8 watts
D. 3.8 watts

A: When the voltage is 13.8 volts DC and the current is 10 amperes, the power is 138 watts. Power is calculated by multiplying voltage by current: 13.8 × 10 = 138.

T5C10. How much power is delivered by a voltage of 12 volts DC and a current of 2.5 amperes?

A. 4.8 watts
B. 30 watts
C. 14.5 watts
D. 0.208 watts

B: When the applied voltage is 12 volts DC and the current is 2.5 amperes, 30 watts of power are generated. The power can be calculated by multiplying the voltage by the current: 12 x 2.5 = 30.

T5C11. How much current is required to deliver 120 watts at a voltage of 12 volts DC?

A. 0.1 amperes
B. 10 amperes
C. 12 amperes
D. 132 amperes

B: When the applied voltage is 12 volts DC and the load is 120 watts, 10 amperes are flowing in a circuit. The current can be calculated by dividing the load by the applied voltage: 120/12 = 10.

T5C12. What is impedance?

A. The opposition to AC current flow
B. The inverse of resistance
C. The inverse of reactance
D. The power handling capability of a component

A: Impedance is the opposition to AC current flow in a circuit. It is analogous to resistance in a DC circuit.

T5D – Ohm's Law; Series and parallel circuits

T5D01. What formula is used to calculate current in a circuit?

A. I = E × R
B. I = E / R
C. I = E2 × R
D. I = E2 / R

B: The formula for calculating current in a circuit is current equals voltage divided by resistance.

T5D02. What formula is used to calculate voltage in a circuit?

A. E = I × R
B. E = I / R
C. E = I2 × R
D. E = I2 / R

A: The formula for calculating voltage in a circuit is voltage equals current multiplied by resistance.

T5D03. What formula is used to calculate resistance in a circuit?

A. R = E × I
B. R = E / I
C. R = E + I
D. R = E - I

B: The formula for calculating resistance in a circuit is resistance (R) equals voltage (E) divided by current (I). Resistance is the opposition to the passage of current, the result of which is the conversion of electrical energy into heat or another type of energy.

T5D04. What is the resistance of a circuit in which a current of 3 amperes flows when connected to 90 volts?

A. 3 ohms
B. 30 ohms
C. 1/30 ohm
D. 270 ohms

B: The resistance of a circuit in which a current of 3 amperes flows when connected to 90 volts is 30 ohms. Resistance is calculated by dividing voltage by current. In this case, then, the resistance is found 90/3 = 30.

T5D05. What is the resistance of a circuit for which the applied voltage is 12 volts and the current flow is 1.5 amperes?

A. 18 ohms
B. 0.125 ohms
C. 8 ohms
D. 13.5 ohms

C: The resistance of a circuit with 12 volts applied and 1.5 amperes of current flow is 8 ohms. Resistance is calculated by dividing voltage by current: 12 ÷ 1.5 = 8.

T5D06. What is the resistance of a circuit that draws 4 amperes from a 12-volt source?

A. 3 ohms
B. 16 ohms
C. 48 ohms
D. 8 ohms

A: The resistance of a circuit that draws 4 amperes from a 12-volt source is 3 ohms. Resistance is calculated by dividing voltage by current: 12/4 = 3.

T5D07. What is the current in a circuit with an applied voltage of 120 volts and a resistance of 80 ohms?

A. 9600 amperes
B. 200 amperes
C. 0.667 amperes
D. 1.5 amperes

D: The current flow in a circuit with an applied voltage of 120 V and a resistance of 80 ohms is 1.5 amperes. The current flow can be found by dividing the voltage by the resistance: 120/80 = 1.5.

T5D08. What is the current through a 100-ohm resistor connected across 200 volts?

A. 20,000 amperes
B. 0.5 amperes
C. 2 amperes
D. 100 amperes

C: The current flowing through a 100-ohm resistor connected across 200 volts is 2 amperes. The current is found by dividing the voltage by the resistance: 200/100 = 2.

T5D09. What is the current through a 24-ohm resistor connected across 240 volts?

A. 2400 amperes
B. 0.1 amperes
C. 10 amperes
D. 5760 amperes

C: The current flowing through a 24-ohm resistor connected across 240 volts is 10 amperes. The current is found by dividing the voltage by the resistance: 240/24 = 10.

T5D10. What is the voltage across a 2-ohm resistor if a current of 0.5 amperes flows through it?

A. 1 volt
B. 0.25 volts
C. 2.5 volts
D. 1.5 volts

A: The voltage across a 2-ohm resistor if a current of 0.5 amperes flows through it is 1 volt. The voltage can be calculated by multiplying the current by the resistance: 2 x 0.5 = 1.

T5D11. What is the voltage across a 10-ohm resistor if a current of 1 ampere flows through it?

A. 1 volt
B. 10 volts
C. 11 volts
D. 9 volts

B: The voltage across a 10-ohm resistor if a current of 1 ampere flows through it is 10 volts. The voltage can be calculated by multiplying the current by the resistance: 10 x 1 = 10.

T5D12. What is the voltage across a 10-ohm resistor if a current of 2 amperes flows through it?

A. 8 volts
B. 0.2 volts
C. 12 volts
D. 20 volts

D: The voltage across a 10-ohm resistor if a current of 2 amperes flows through it is 20 volts. The voltage can be calculated by multiplying the current by the resistance: 10 x 2 = 20.

T5D13. In which type of circuit is DC current the same through all components?

A. Series
B. Parallel
C. Resonant
D. Branch

A: Components in series will have the same current flowing through them. However, the voltage will not be the same for each.

T5D14. In which type of circuit is the voltage always the same across all components?

A. Series
B. Parallel
C. Resonant
D. Branch

B: Components in parallel will have the same voltage across them. However, the current will not be the same for each.

T6 – Electronic and Electrical Components

T6A – Fixed and Variable Resistors; Capacitors; Inductors; Fuses; Switches; Batteries

T6A01. What electrical component opposes the flow of current in a DC circuit?

A. Inductor
B. Resistor
C. Inverter
D. Transformer

B: The electrical component used to oppose the flow of current in a DC circuit is a resistor. For ham radio technicians, the most commonly required resistors are quarter- and half-watt fixed value resistors and 100, 500, 1k, 5k, 10k, and 100k ohm variable resistors.

T6A02. What type of component is often used as an adjustable volume control?

A. Fixed resistor
B. Power resistor
C. Potentiometer
D. Transformer

C: A potentiometer is often used as an adjustable volume control. A potentiometer is an instrument that measures an unknown voltage by comparing it to a standard voltage.

T6A03. What electrical parameter is controlled by a potentiometer?

A. Inductance
B. Resistance
C. Capacitance
D. Field strength

B: Resistance is controlled by a potentiometer. A potentiometer will typically have a resistance element connected to the circuit with three terminals. Two of these are attached to the two input voltage conductors of the circuit, and the third is connected to the circuit output. The third terminal is typically able to move across the resistance element, which enables the magnitude of the input voltage to be adjusted.

T6A04. What electrical component stores energy in an electric field?

A. Resistor
B. Capacitor
C. Inductor
D. Diode

B: The capacitor is the electrical component that stores energy in an electric field. Typically, a capacitor is composed of two metallic plates that are separated and insulated from one another by a dielectric.

T6A05. What type of electrical component consists of conductive surfaces separated by an insulator?

A. Resistor
B. Potentiometer
C. Oscillator
D. Capacitor

D: A capacitor consists of two or more conductive surfaces separated by an insulator.

T6A06. What type of electrical component stores energy in a magnetic field?

A. Resistor
B. Capacitor
C. Inductor
D. Diode

C: An inductor stores energy in a magnetic field. Inductors are typically composed of at least one wire loop. An inductor may be fixed, adjustable, or variable, depending on whether the number of loops is able to be manipulated.

T6A07. What electrical component is typically constructed as a coil of wire?

A. Transistor
B. Capacitor
C. Diode
D. Inductor

D: An inductor is usually composed of a coil of wire.

T6A08. What is the function of an SPDT switch?

A. A single circuit is opened or closed
B. Two circuits are opened or closed
C. A single circuit is switched between one of two other circuits
D. Two circuits are each switched between one of two other circuits

C: An SPDT switch can toggle a single circuit between one of two other circuits. SPDT stands for Single Pole Double Throw, which means it has a single input and two possible outputs.

T6A09. What type of switch is represented by component 3 in figure T-2?

A. Single-pole single-throw
B. Single-pole double-throw
C. Double-pole single-throw
D. Double-pole double-throw

A: A single-pole single-throw switch is represented by component 3 in figure T-2.

T6A10. Which of the following battery chemistries is rechargeable?

A. Nickel-metal hydride
B. Lithium-ion
C. Lead-acid
D. All these choices are correct

D: There are many different types of batteries that may be recharged, including nickel-metal hydride (NiMH), lithium-ion (Li-ion), and lead-acid.

T6A11. Which of the following battery chemistries is not rechargeable?

A. Nickel-cadmium
B. Carbon-zinc
C. Lead-acid
D. Lithium-ion

B: The carbon-zinc battery is not rechargeable.

T6B – Semiconductors: Basic Principles and Applications of Solid-State Devices, Diodes and Transistors; Gain

T6B01. Which is true about forward voltage drop in a diode?

A. It is lower in some diode types than in others
B. It is proportional to peak inverse voltage
C. It indicates that the diode is defective
D. It has no impact on the voltage delivered to the load

A: Forward voltage drop is lower in some diode types than in others.

T6B02. What electronic component allows current to flow in only one direction?

A. Resistor
B. Fuse
C. Diode
D. Driven element

C: A diode allows current to flow in only one direction. A diode is an electronic device that includes two electrodes, an anode and a cathode. These devices can be used to convert an alternating current into a direct current.

T6B03. Which of these components can be used as an electronic switch?

A. Varistor
B. Potentiometer
C. Transistor
D. Thermistor

C: A transistor can be used as an electronic switch. It may also be used as a detector.

T6B04. Which of the following components can consist of three regions of semiconductor material?

A. Alternator
B. Transistor
C. Triode
D. Pentode

B: A transistor is made of three regions of semiconductor material.

T6B05. What type of transistor has a gate, drain, and source?

A. Varistor
B. Field-effect
C. Hall-effect
D. Bipolar junction

B: Field-effect transistors have three pins called gate, drain, and source.

T6B06. How is the cathode lead of a semiconductor diode often marked on the package?

A. With the word "cathode"
B. With a stripe
C. With the letter C
D. With the letter K

B: A semiconductor diode's cathode lead is usually identified with a stripe.

T6B07. What causes a light-emitting diode (LED) to emit light?

A. Forward current
B. Reverse current
C. Capacitively-coupled RF signal
D. Inductively-coupled RF signal

A: An LED is a semiconductor diode that, when a forward current passes through it, produces light in or near the visible range. Each diode is only capable of producing a single primary color, though diodes may be combined to produce any color.

T6B08. What does the abbreviation FET stand for?

A. Frequency Emission Transmitter
B. Fast Electron Transistor
C. Free Electron Transmitter
D. Field Effect Transistor

D: The abbreviation FET stands for Field Effect Transistor. In a field effect transistor, output current is controlled by a variable electric field.

T6B09. What are the names for the electrodes of a diode?

A. Plus and minus
B. Source and drain
C. Anode and cathode
D. Gate and base

C: The names of the two electrodes of a diode are the anode and cathode. Electrons leave a system through the anode and return through the cathode.

T6B10. Which of the following can provide power gain?

A. Transformer
B. Transistor
C. Reactor
D. Resistor

B: Transformers don't provide a gain. From the options provided, only a transistor provides a gain.

T6B11. What does the term gain mean in amplifiers?

A. The output signal voltage relative to the input signal voltage
B. The output signal current relative to the input signal current
C. The output signal power relative to the input signal power
D. All these choices are correct

D: In amplifiers, gain means the output signal relative to the input signal, whether measured as voltage, current, or power.

T6B12. What are the names of the electrodes of a bipolar junction transistor?

A. Signal, bias, power
B. Emitter, base, collector
C. Input, output, supply
D. Pole one, pole two, output

B: The three electrodes of a bipolar junction transistor are the emitter, the base, and the collector, often abbreviated E, B, and C, respectively.

T6C – Circuit Diagrams: Use of Term Schematic, Basic Structure; Schematic Symbols of Basic Components

T6C01. What is an electrical diagram using standard component symbols called?

A. Connection chart
B. Instrumentation system
C. Schematic
D. Flow chart

C: An electrical diagram with standard wiring component symbols is called a schematic.

T6C02. What is component 1 in the figure below?

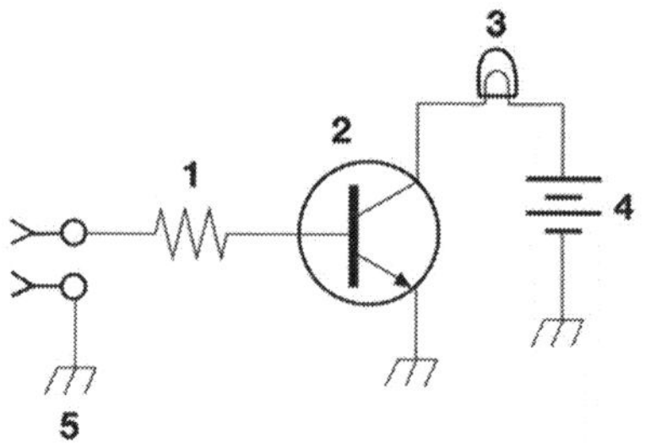

A. Resistor
B. Transistor
C. Battery
D. Connector

A: In the figure, component 1 is a resistor.

T6C03. What is component 2 in the figure below?

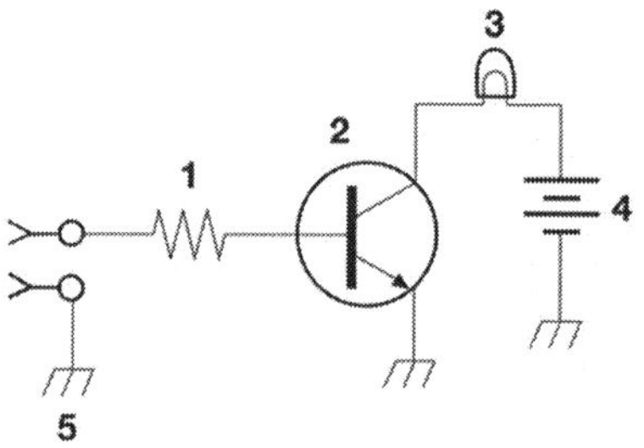

A. Resistor
B. Transistor
C. Indicator lamp
D. Connector

B: In the figure, component 2 is a transistor.

T6C04. What is component 3 in the figure below?

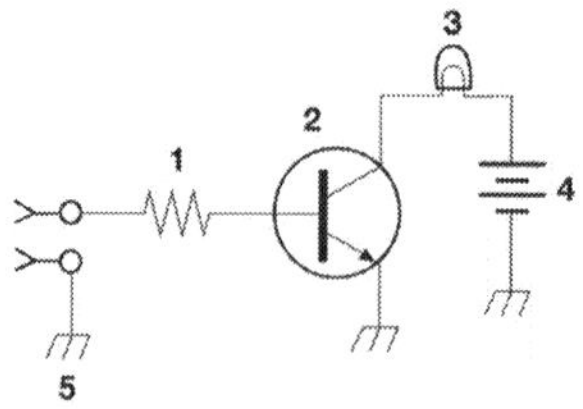

A. Resistor
B. Transistor
C. Lamp
D. Ground symbol

C: In the figure, component 3 is a lamp.

T6C05. What is component 4 in the figure below?

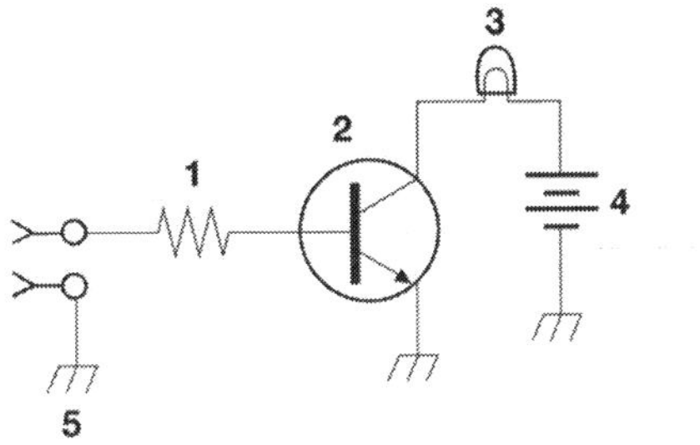

A. Resistor
B. Transistor
C. Ground symbol
D. Battery

D: In the figure, component 4 is a battery.

T6C06. What is component 6 in the figure below?

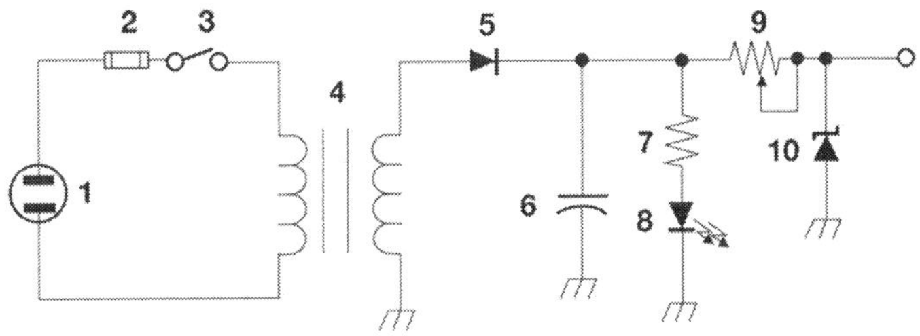

A. Resistor
B. Capacitor
C. Regulator IC
D. Transistor

B: In the figure, component 6 is a capacitor.

T6C07. What is component 8 in the figure below?

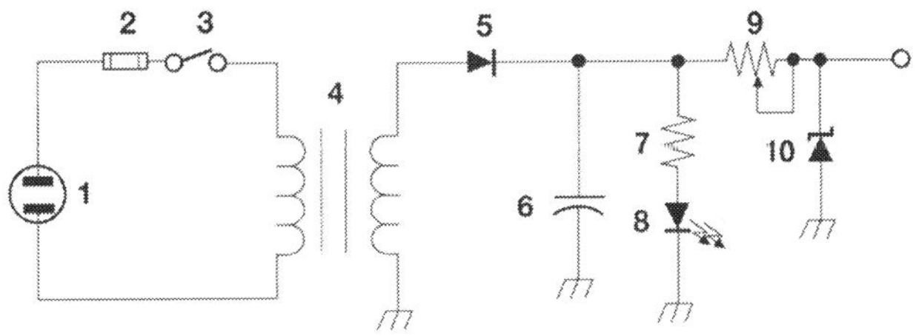

A. Resistor
B. Inductor
C. Regulator IC
D. Light emitting diode

D: In the figure, component 8 is a light emitting diode.

T6C08. What is component 9 in the figure below?

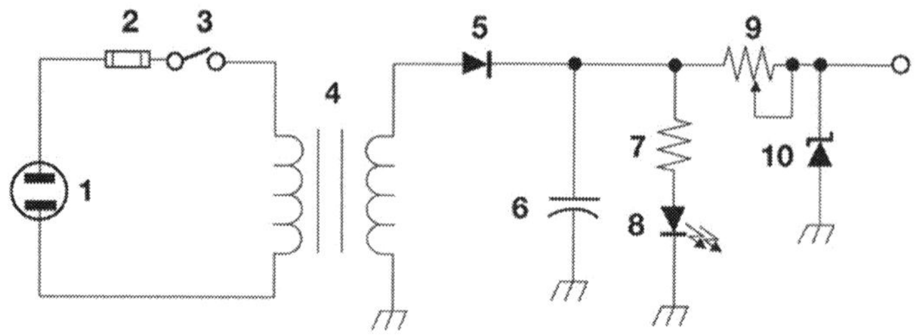

A. Variable capacitor
B. Variable inductor
C. Variable resistor
D. Variable transformer

C: In the figure, component 9 is a variable resistor.

T6C09. What is component 4 in the figure below?

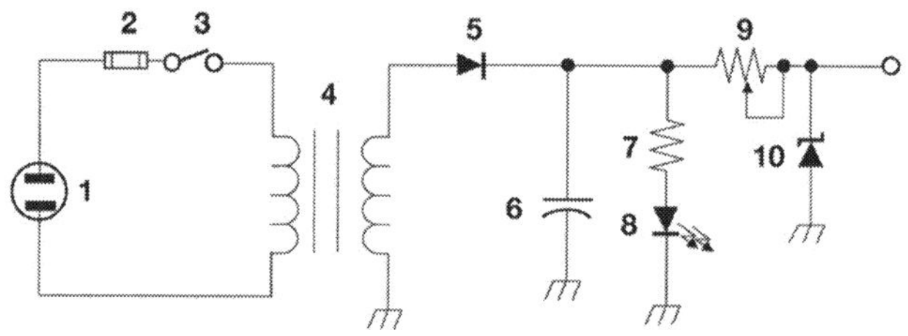

A. Variable inductor
B. Double-pole switch
C. Potentiometer
D. Transformer

D: In the figure, component 4 is a transformer.

T6C10. What is component 3 in the figure below?

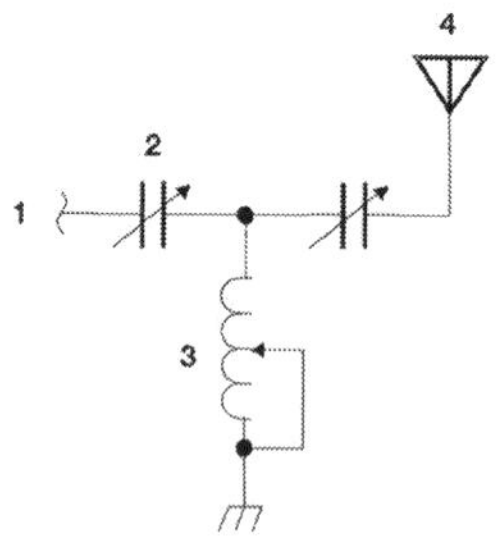

A. Connector
B. Meter
C. Variable capacitor
D. Variable inductor

D: In the figure, component 3 is a variable inductor.

T6C11. What is component 4 in the figure above?

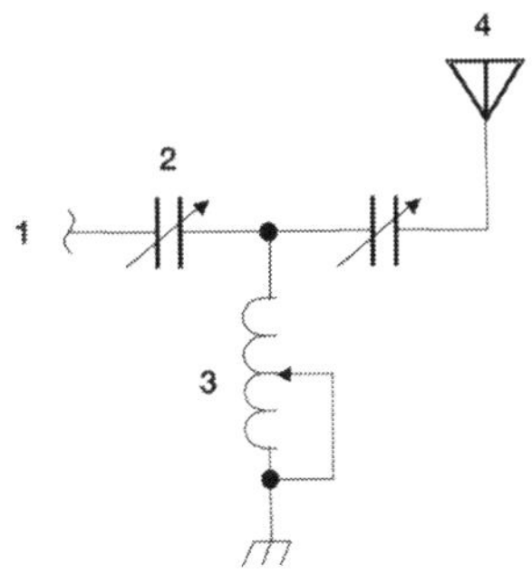

A. Antenna
B. Transmitter
C. Dummy load
D. Ground

A: In the figure, component 4 is an antenna.

T6C12. Which of the following is accurately represented in electrical schematics?

A. Wire lengths
B. Physical appearance of components
C. Component connections
D. All these choices are correct

C: The connections on an electrical circuit schematic diagram represent the actual connections of electrical components.

T6D – Component Functions: Rectifiers, Relays, Voltage Regulators, Meters, Indicators, Integrated Circuits, Transformers; Resonant Circuit; Shielding

T6D01. Which of the following devices or circuits changes an alternating current into a varying direct current signal?

A. Transformer
B. Rectifier
C. Amplifier
D. Reflector

B: A rectifier changes an alternating current into a varying direct current signal. Specifically, it is a switch that blocks the current in one direction but permits it in the other.

T6D02. What is a relay?

A. An electrically-controlled switch
B. A current-controlled amplifier
C. An inverting amplifier
D. A pass transistor

A: A relay may best be described as a switch controlled by an electromagnet. When the device senses a miniscule voltage or current change, it activates a switch or another component in the electric circuit.

T6D03. Which of the following is a reason to use shielded wire?

A. To decrease the resistance of DC power connections
B. To increase the current carrying capability of the wire
C. To prevent coupling of unwanted signals to or from the wire
D. To reduce receiver overload

C: Shielding a wire is necessary to prevent coupling of unwanted signals to or from the wire.

T6D04. Which of the following displays an electrical quantity as a numeric value?

A. Potentiometer
B. Transistor
C. Meter
D. Relay

C: A meter is a device used to display an electrical quantity as a numeric value. Meters can be used to show measurements such as voltage, current, or resistance.

T6D05. What type of circuit controls the amount of voltage from a power supply?

A. Regulator
B. Oscillator
C. Filter
D. Phase inverter

A: A regulator circuit controls the amount of voltage from a power supply.

T6D06. What component changes 120 V AC power to a lower AC voltage for other uses?

A. Variable capacitor
B. Transformer
C. Transistor
D. Diode

B: A transformer is commonly used to change 120V AC house current to a lower AC voltage for other uses.

T6D07. Which of the following is commonly used as a visual indicator?

A. LED
B. FET
C. Zener diode
D. All these choices are correct

A: LED is commonly used as a visual indicator.

T6D08. Which of the following is combined with an inductor to make a resonant circuit?

A. Resistor
B. Zener diode
C. Potentiometer
D. Capacitor

D: A capacitor is used along with an inductor to make a resonant circuit.

T6D09. What is the name of a device that combines several semiconductors and other components into one package?

A. Transducer
B. Multi-pole relay
C. Integrated circuit
D. Transformer

C: An integrated circuit combines several semiconductors and other components into one package. All of these semiconductors and other components are either imprinted or etched on a thin slice of a semiconducting material, as for instance silicon.

T6D10. What is the function of component 2 in the figure below?

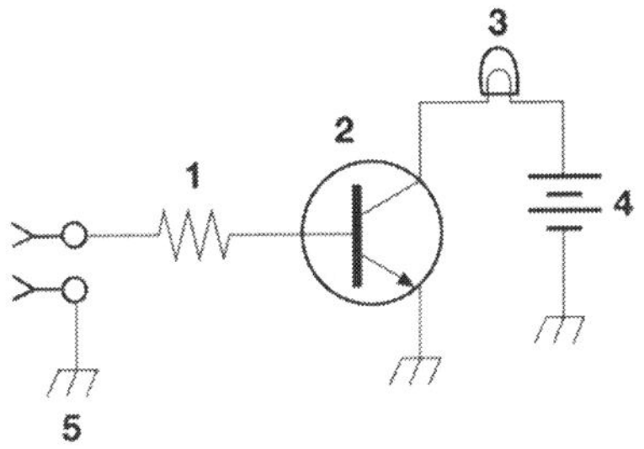

A. Give off light when current flows through it
B. Supply electrical energy
C. Control the flow of current
D. Convert electrical energy into radio waves

C: In the figure, the function of component 2 is to control the flow of current.

T6D11. Which of the following is a resonant or tuned circuit?

A. An inductor and a capacitor in series or parallel
B. A linear voltage regulator
C. A resistor circuit used for reducing standing wave ratio
D. A circuit designed to provide high-fidelity audio

A: A simple resonant or tuned circuit is a circuit containing an inductor and a capacitor, connected either in series or in parallel, which functions to filter out certain frequencies.

T7 – Practical Circuits

T7A – Station Equipment: Receivers, Transceivers, Transmitter Amplifiers, RF Preamplifiers, Transverters; Basic Radio Circuit Concepts and Terminology: Sensitivity, Selectivity, Mixers, Oscillators, Push-To-Talk (PTT), VFO, Modulation

T7A01. Which term describes the ability of a receiver to detect the presence of a signal?

A. RF gain
B. Sensitivity
C. Selectivity
D. Total Harmonic Distortion

B: The ability of a receiver to detect the presence of a signal is measured in terms of its sensitivity.

T7A02. What is a transceiver?

A. A device that combines a receiver and transmitter
B. A device for matching feed line impedance to 50 ohms
C. A device for automatically sending and decoding Morse code
D. A device for converting receiver and transmitter frequencies to another band

A: A transceiver is a device that combines the functions of a transmitter and a receiver.

T7A03. Which of the following is used to convert a signal from one frequency to another?

A. Phase splitter
B. Mixer
C. Inverter
D. Amplifier

B: A mixer is a device used to convert radio signals from one frequency to another.

T7A04. Which term describes the ability of a receiver to discriminate between multiple signals?

A. Discrimination ratio
B. Sensitivity
C. Selectivity
D. Harmonic distortion

C: A receiver's ability to discriminate among multiple radio signals is referred to as its selectivity.

T7A05. What is the name of a circuit that generates a signal at a specific frequency?

A. Reactance modulator
B. Phase modulator
C. Low-pass filter
D. Oscillator

D: An oscillator is a circuit that produces an oscillating signal at a desired frequency.

T7A06. What device converts the RF input and output of a transceiver to another band?

A. High-pass filter
B. Low-pass filter
C. Transverter
D. Phase converter

C: A transverter takes the output of a low-powered 28 MHz SSB exciter and produces a 222 MHz output signal. Transverters are a single unit that includes both an upconverter and a downconverter. The transverter enables the transceiver to communicate over a broad range of frequencies.

T7A07. What is the function of a transceiver's PTT input?

A. Input for a key used to send CW
B. Switches transceiver from receive to transmit when grounded
C. Provides a transmit tuning tone when grounded
D. Input for a preamplifier tuning tone

B: The term PTT stands for push-to-talk, a setup in which the transceiver is set to receive when the button is not pushed, and transmit when the button is pushed (when the switch is grounded).

T7A08. Which of the following describes combining speech with an RF carrier signal?

A. Impedance matching
B. Oscillation
C. Modulation
D. Low-pass filtering

C: The process of combining speech with an RF carrier signal is referred to as modulation.

T7A09. What is the function of the switch which selects either SSB or CW-FM on some VHF power amplifiers?

A. Change the mode of the transmitted signal
B. Set the amplifier for proper operation in the selected mode
C. Change the frequency range of the amplifier to operate in the proper segment of the band
D. Reduce the received signal noise

B: Single Side Band/Continuous Wave - FM switch on a VHF amplifier sets the amplifier to the selected mode. The switch provides an easy toggle between settings.

T7A10. What can be added to the output of a transceiver to increase the transmitted output power?

A. A potentiometer
B. An RF power amplifier
C. An impedance multiplier
D. All these choices are correct

B: An RF power amplifier increases the low-power output from a handheld transceiver.

T7A11. What is the function of the Variable Frequency Oscillator (VFO) circuit in a transceiver?

A. Set the receive and transmit frequency
B. Provide automatic frequency control
C. Inject a variable frequency to allow CW reception
D. Generate and demodulate single sideband signals

A: The Variable Frequency Oscillator (VFO) circuit in a transceiver sets the receive and transmit frequency.

T7B – Symptoms, Causes, and Cures of Common Transmitter and Receiver Problems: Overload and Overdrive, Distortion, Interference and Consumer Electronics, RF Feedback

T7B01. What can you do if you are told your FM handheld or mobile transceiver is over-deviating?

A. Talk louder into the microphone
B. Let the transceiver cool off
C. Change to a higher power level
D. Talk farther away from the microphone

D: If you are told your FM handheld or mobile transceiver is over deviating, you can talk farther away from the microphone.

T7B02. What would cause a broadcast AM or FM radio to receive an amateur radio transmission unintentionally?

A. The receiver is unable to reject strong signals outside the AM or FM band
B. The microphone gain of the transmitter is turned up too high
C. The audio amplifier of the transmitter is overloaded
D. The deviation of an FM transmitter is set too low

A: If a broadcast AM/FM radio is unintentionally receiving amateur radio transmissions, the most likely explanation is that the receiver is unable to reject strong signals outside of the AM/FM bands.

T7B03. Which of the following can cause radio frequency interference?

A. Fundamental overload
B. Harmonics
C. Spurious emissions
D. All these choices are correct

D: Radio frequency interference may be caused by fundamental overload, harmonics, or spurious emissions. A spurious emission is any radio frequency that was not created or transmitted on purpose.

T7B04. Which of the following might be the cause of low RF power output from a solid-state transceiver?

A. Poor amplifier noise figure
B. Poor amplifier linearity
C. Low SWR
D. High SWR

D: High SWR might be the cause of low RF power output from a solid-state transceiver.

T7B05. Which of the following might reduce interference by an amateur station to a non-amateur over-the-air radio receiver?

A. Block the amateur signal with a filter at the antenna input of the affected receiver
B. Block the interfering signal with a filter on the amateur transmitter
C. Switch the transmitter from FM to SSB
D. Switch the transmitter to a narrow-band mode

A: Filtering out an unwanted amateur radio signal at the receiver of the impacted device is the best way to reduce or eliminate overload on a TV or radio.

T7B06. Which of the following actions should you take if a neighbor tells you that your station's transmissions are interfering with their radio or TV reception?

A. Make sure that your station is functioning properly and that it does not cause interference to your own radio or television when it is tuned to the same channel
B. Immediately turn off your transmitter and contact the nearest FCC office for assistance
C. Install a harmonic doubler on the output of your transmitter and tune it until the interference is eliminated
D. All these choices are correct

A: If someone tells you that your station's transmissions are interfering with their radio or TV reception, your first step should be to make sure that your station is functioning properly and that it does not cause interference to your own television.

T7B07. Which of the following can reduce interference to a 2-meter band transceiver from a nearby commercial FM station?

A. Installing an RF preamplifier
B. Using double-shielded coaxial cable
C. Installing bypass capacitors on the microphone cable
D. Installing a band-reject filter

D: The purpose of a band reject filter is specifically for blocking specific signals, especially if they are strong, to prevent signal overload.

T7B08. What should you do if something in a neighbor's home is causing harmful interference to your amateur station?

A. Work with your neighbor to identify the offending device
B. Politely inform your neighbor that FCC rules prohibit the use of devices that cause interference
C. Make sure your station meets the standards of good amateur practice
D. All these choices are correct

D: If you believe something in your neighbor's house is causing harmful interference to your amateur station, there are a few things you can or should do. First, make sure that your station is set up properly and that a faulty setup is not causing the interference problems. If that is not successful, talk to your neighbor about the interference problems you are experiencing, and politely let him know about the FCC rules that govern harmful interference. If your neighbor is amenable, work with him to locate the source of the interference and find a solution. Always remain courteous.

T7B09. What should be the first step to resolve non-fiber optic cable TV interference caused by your amateur radio transmission?

A. Add a low-pass filter to the TV antenna input
B. Add a high-pass filter to the TV antenna input
C. Add a preamplifier to the TV antenna input
D. Be sure all TV feed line coaxial connectors are installed properly

D: To remedy any non-fiber optic cable TV interference caused by your amateur radio transmission, ensure that all TV feed line coaxial connectors are installed properly.

T7B10. What might be a problem if you receive a report that your audio signal through an FM repeater is distorted or unintelligible?

A. Your transmitter is slightly off frequency
B. You are speaking too loudly or too close to the microphone
C. You are in a bad location
D. All these choices are correct

D: If you receive a report that your audio signal through the repeater is distorted or unintelligible, this may be because your transmitter is slightly off frequency, your batteries are running low, or you're in a bad location.

T7B11. Which of the following can eliminate distorted voice transmissions?

A. Adding extra feedline to the antenna to lower SWR
B. Turning the radio on and off to reset the computer-controlled circuitry
C. Adding a clip-on ferrite "choke" to the microphone cable to prevent the transmitted signal from feeding back into the transmitter
D. Turning the squelch control fully clockwise to prevent the transmitted signal from triggering the squelch circuit

C: Adding a clip-on ferrite choke to the microphone cable can eliminate distorted voice transmissions by preventing transmitted RF from feeding back into the transmitter.

T7C – Antenna and Transmission Line Measurements and Troubleshooting: Measuring SWR, Effects of High SWR, Causes of Feed Line Failures; Basic Coaxial Cable Characteristics; Use of Dummy Loads When Testing

T7C01. What is the primary purpose of a dummy load?

A. To prevent transmitting signals over the air when making tests
B. To prevent over-modulation of a transmitter
C. To improve the efficiency of an antenna
D. To improve the signal-to-noise ratio of a receiver

A: The primary purpose of a dummy load is to prevent the radiation of signals when making tests. The dummy load, also known as a dummy antenna, is a station accessory that enables the user to test and finetune equipment without actually broadcasting a signal.

T7C02. Which of the following is used to determine if an antenna is resonant at the desired operating frequency?

A. A VTVM
B. An antenna analyzer
C. A Q meter
D. A frequency counter

B: An antenna analyzer can be used to determine if an antenna is resonant at the desired operating frequency. Antenna analyzers assess the efficiency of antenna systems.

T7C03. What does a typical RF dummy load consist of?

A. A low-voltage power supply and an AC relay
B. A 50-ohm non-inductive resistor mounted on a heat sink
C. A low-voltage power supply and a DC relay
D. A 50-ohm inductive reactance mounted in a shielded enclosure

B: A typical RF dummy load consists of a 50-ohm non-inductive resistor mounted on a heat sink. It is designed to absorb transmitter power without radiating a signal.

T7C04. What reading on an SWR meter indicates a perfect impedance match between the antenna and the feed line?

A. 50:50
B. Zero
C. 1:1
D. Full Scale

C: An SWR meter reading of 1:1 indicates a perfect impedance match between the antenna and the feed line. This means that all of the transmitted power is being delivered to the load with no reflected power.

T7C05. Why do most solid-state transmitters reduce output power as SWR increases beyond a certain level?

A. To protect the RF output amplifier transistors
B. To comply with FCC rules on spectral purity
C. Because power supplies cannot supply enough current at high SWR
D. To lower the SWR on the transmission line

A: Solid-state radios will automatically reduce the power output in order to prevent damage, specifically to output amplifier transistors, from an increase in SWR.

T7C06. What does an SWR reading of 4:1 indicate?

A. Loss of -4 dB
B. Good impedance match
C. Gain of +4 dB
D. Impedance mismatch

D: An SWR reading of 4:1 indicates an impedance mismatch.

T7C07. What happens to power lost in a feed line?

A. It increases the SWR
B. It is radiated as harmonics
C. It is converted into heat
D. It distorts the signal

C: The power lost in a feed line is converted into heat. If the cable has cracks or dents, it will be more likely to lose power in this manner. A ham radio operator should always use the best feed line he or she can afford.

T7C08. Which instrument can be used to determine SWR?

A. Voltmeter
B. Ohmmeter
C. Iambic pentameter
D. Directional wattmeter

D: Besides an SWR meter, one could also use a directional wattmeter to determine if a feed line and antenna are properly matched. An SWR (standing wave ratio) meter can evaluate the effectiveness of impedance matching, and can also identify the degree of mismatch between a transmission line and its load. A directional wattmeter is capable of differentiating between the forward and reflected waves.

T7C09. Which of the following causes failure of coaxial cables?

A. Moisture contamination
B. Solder flux contamination
C. Rapid fluctuation in transmitter output power
D. Operation at 100% duty cycle for an extended period

A: The most common cause of failure in coaxial cables is moisture contamination. Coaxial cable, also known as coax, is composed of a central wire surrounded by hollow tube. The center wire is called the center conductor, and the outer conductor is known as the braid, or shield.

T7C10. Why should the outer jacket of coaxial cable be resistant to ultraviolet light?

A. Ultraviolet light can increase the resistance of the conductors
B. Ultraviolet light can increase losses in the cable's jacket
C. Ultraviolet and RF signals can mix, causing interference
D. Ultraviolet light can damage the jacket and allow water to enter the cable

D: The outer jacket of a coaxial cable should be resistant to ultraviolet light because this light can damage the jacket and allow water to enter the cable.

T7C11. What is an advantage of foam-dielectric versus solid-dielectric coaxial cable?

A. It is more resistant to moisture contamination
B. It has higher voltage breakdown
C. It has less loss per foot
D. It has a better impedance match to 50 ohms

C: Foam-dielectric coaxial cable has an advantage over solid-dielectric coaxial cable because it has less loss per foot.

T7D – Using Basic Test Instruments: Voltmeter, Ammeter, and Ohmmeter; Soldering

T7D01. Which instrument would you use to measure electric potential?

A. An ammeter
B. A voltmeter
C. A potentiometer
D. An ohmmeter

B: A voltmeter is used to measure electric potential or electromotive force. It accomplishes this by identifying the electric potential difference between two conductors.

T7D02. How is a voltmeter connected to a component to measure applied voltage?

A. In series
B. In parallel
C. In quadrature
D. In phase

B: The correct way to connect a voltmeter to a circuit is in parallel with the circuit. An analog voltmeter, on the other hand, is a moving coil galvanometer that works as a voltmeter after a resistor is installed in series with it.

T7D03. When configured to measure current, how is a multimeter connected to a component?

A. In series
B. In parallel
C. In quadrature
D. In phase

A: An ammeter is usually connected in series with the circuit. In most cases, an ammeter is combined with a voltmeter and ohmmeter in a single instrument.

T7D04. Which instrument is used to measure electric current?

A. An ohmmeter
B. An electrometer
C. A voltmeter
D. An ammeter

D: An ammeter is used to measure electric current. Typically, the current will need to be at least a few amperes to be susceptible to measurement by the ammeter.

T7D05. How does an ohmmeter measure the resistance of a circuit or component?

A. By applying a small current and measuring the resulting voltage
B. By placing a variable resistor in parallel with the circuit
C. By placing a variable resistor in series with the circuit
D. By applying a variable voltage and measuring the resulting current change

A: An ohmmeter measures resistance by applying a small current and measuring the resulting voltage.

T7D06. Which of the following can damage a multimeter?

A. Attempting to measure resistance using the voltage setting
B. Failing to connect one of the probes to ground
C. Attempting to measure voltage when using the resistance setting
D. Not allowing it to warm up properly

C: Attempting to measure voltage when using the resistance setting might damage a multimeter. Multimeters are capable of measuring the following values: current, resistance, and the current potential differences of alternating and direct current.

T7D07. Which of the following measurements are made using a multimeter?

A. Signal strength and noise
B. Impedance and reactance
C. Voltage and resistance
D. All these choices are correct

C: Voltage and resistance measurements are commonly made using a multimeter. The multimeter is sometimes referred to as a volt-ohm meter.

T7D08. Which of the following types of solder should not be used for radio and electronic applications?

A. Acid-core solder
B. Lead-tin solder
C. Rosin-core solder
D. Tin-copper solder

A: Rosin-core solder is best for radio and electronic use. Acid-core should not be used.

T7D09. What is the characteristic appearance of a cold tin-lead solder joint?

A. Dark black spots
B. A bright or shiny surface
C. A rough or lumpy surface
D. A greenish tinge

C: Cold tin-lead solder joints typically have a rough or lumpy surface.

T7D10. What reading indicates that an ohmmeter is connected across a large, discharged capacitor?

A. Increasing resistance with time
B. Decreasing resistance with time
C. Steady full-scale reading
D. Alternating between open and short circuit

A: When an ohmmeter is connected across a large, discharged capacitor it will show increasing resistance over time.

T7D11. Which of the following precautions should be taken when measuring in-circuit resistance with an ohmmeter?

A. Ensure that the applied voltages are correct
B. Ensure that the circuit is not powered
C. Ensure that the circuit is grounded
D. Ensure that the circuit is operating at the correct frequency

B: When measuring in-circuit resistance with an ohmmeter, one should ensure that the circuit is not powered.

T8 – SIGNALS AND EMISSIONS

T8A – BASIC CHARACTERISTICS OF FM AND SSB; BANDWIDTH OF VARIOUS MODULATION MODES: CW, SSB, FM, FAST-SCAN TV; CHOICE OF EMISSION TYPE: SELECTION OF USB VS LSB, USE OF SSB FOR WEAK SIGNAL WORK, USE OF FM FOR VHF PACKET AND REPEATERS

T8A01. Which of the following is a form of amplitude modulation?

A. Spread spectrum
B. Packet radio
C. Single sideband
D. Phase shift keying (PSK)

C: Single sideband is a form of amplitude modulation. Indeed, single sideband is the most common mode of voice transmission on the high-frequency bands. Unlike an a.m. transmission, in which two identical versions of the message are broadcast, an SSB transmission only broadcasts one. For this reason, single sideband communications are more efficient and are better at conserving space on the radio spectrum.

T8A02. What type of modulation is commonly used for VHF packet radio transmissions?

A. FM or PM
B. SSB
C. AM
D. PSK

A: The type of modulation most commonly used for VHF packet radio transmission is either FM or PM.

T8A03. Which type of voice mode is often used for long-distance (weak signal) contacts on the VHF and UHF bands?

A. FM
B. DRM
C. SSB
D. PM

C: The type of voice modulation often used for long-distance or weak signal contacts on the VHF and UHF bands is SSB.

T8A04. Which type of modulation is commonly used for VHF and UHF voice repeaters?

A. AM
B. SSB
C. PSK
D. FM or PM

D: The type of modulation most commonly used for VHF and UHF voice repeaters is either FM or PM.

T8A05. Which of the following signal types has the narrowest bandwidth?

A. FM voice
B. SSB voice
C. CW
D. Slow-scan TV

C: CW has a narrower bandwidth than FM voice, SSB voice, or slow-scan TV emission.

T8A06. Which sideband is normally used for 10-meter HF, VHF, and UHF single-sideband communications?

A. Upper sideband
B. Lower sideband
C. Suppressed sideband
D. Inverted sideband

A: The upper sideband is normally used for 10 meter HF, VHF, and UHF single-sideband communications.

T8A07. What is one characteristic of single sideband (SSB) compared to FM?

A. SSB signals are easier to tune in correctly
B. SSB signals are less susceptible to interference
C. SSB signals have narrower bandwidth
D. SSB signals are less susceptible to high SWR

C: The primary advantage of single sideband over FM is that SSB signals have narrower bandwidth.

T8A08. What is the approximate bandwidth of a typical single sideband (SSB) voice signal?

A. 1 kHz
B. 3 kHz
C. 6 kHz
D. 15 kHz

B: The approximate bandwidth of a single sideband voice signal is 3 kHz.

T8A09. What is the approximate bandwidth of an FM voice signal on VHF repeaters?

A. Less than 500 Hz
B. About 150 kHz
C. Between 10 and 15 kHz
D. Between 50 and 125 kHz

C: The approximate bandwidth of a VHF repeater FM voice signal is between 10 and 15 kHz.

T8A10. What is the approximate bandwidth of AM fast-scan TV transmissions?

A. More than 10 MHz
B. About 6 MHz
C. About 3 MHz
D. About 1 MHz

B: The typical bandwidth of AM fast-scan TV transmissions on the 70 cm band is about 6 MHz.

T8A11. What is the approximate bandwidth required to transmit a CW signal?

A. 2.4 kHz
B. 150 Hz
C. 1000 Hz
D. 15 kHz

B: The approximate bandwidth required to transmit a CW signal is 150 Hz.

T8A12. Which of the following is a disadvantage of FM compared with single sideband?

A. Voice quality is poorer
B. Only one signal can be received at a time
C. FM signals are harder to tune
D. FM signals are more susceptible to high SWR

B: A disadvantage of FM when compared to single sideband is that only one signal can be received at a time.

T8B – Amateur Satellite Operation: Doppler Shift, Basic Orbits, Operating Protocols, Modulation Mode Selection, Transmitter Power Considerations, Telemetry, Satellite Tracking Programs, Beacons, Uplink and Downlink Mode Definitions, Spin Fading, Definition of "LEO", Setting Uplink Power

T8B01. What telemetry information is typically transmitted by satellite beacons?

A. The signal strength of received signals
B. Time of day accurate to plus or minus 1/10 second
C. Health and status of the satellite
D. All these choices are correct

C: Health and status are the most common telemetry datapoints that are transmitted by satellite beacons.

T8B02. What is the impact of using excessive effective radiated power on a satellite uplink?

A. Possibility of commanding the satellite to an improper mode
B. Blocking access by other users
C. Overloading the satellite batteries
D. Possibility of rebooting the satellite control computer

B: The result can be that the signal-to-noise ratio received by the repeater can be overpowered by a single signal and thus won't transmit other valid, but weaker signals.

T8B03. Which of the following are provided by satellite tracking programs?

A. Maps showing the real-time position of the satellite track over Earth
B. The time, azimuth, and elevation of the start, maximum altitude, and end of a pass
C. The apparent frequency of the satellite transmission, including effects of Doppler shift
D. All these choices are correct

D: Satellite tracking programs provide a wealth of information about the path and position of orbiting satellites, as well as their operation.

T8B04. What mode of transmission is commonly used by amateur radio satellites?

A. SSB
B. FM
C. CW/data
D. All these choices are correct

D: The three most common transmission modes are SSB, FM, and CW/data.

T8B05. What is a satellite beacon?

A. The primary transmit antenna on the satellite
B. An indicator light that shows where to point your antenna
C. A reflective surface on the satellite
D. A transmission from a satellite that contains status information

D: A satellite beacon is a transmission from a space station that contains information about a satellite. These transmissions are used for maintenance and guidance.

T8B06. Which of the following are inputs to a satellite tracking program?

A. The satellite transmitted power
B. The Keplerian elements
C. The last observed time of zero Doppler shift
D. All these choices are correct

B: The six Keplerian elements precisely define the orbit of a satellite. The six elements are: (1) eccentricity, (2) semi-major axis, (3) inclination, (4) longitude of the ascending node, (5) argument of periapsis, and (6) mean anomaly at epoch.

T8B07. What is Doppler shift in reference to satellite communications?

A. A change in the satellite orbit
B. A mode where the satellite receives signals on one band and transmits on another
C. An observed change in signal frequency caused by relative motion between the satellite and Earth station
D. A special digital communications mode for some satellites

C: With regard to satellite communications, Doppler shift is an observed change in signal frequency caused by relative motion between the satellite and the earth station. When the satellite is approaching the earth station, the wavelength will be shorter; as the satellite moves away from the earth station, the wave length will gradually expand.

T8B08. What does it mean if a satellite is operating in U/V mode?

A. The satellite uplink is in the 15-meter band and the downlink is in the 10 meter band
B. The satellite uplink is in the 70-centimeter band and the downlink is in the 2-meter band
C. The satellite operates using ultraviolet frequencies
D. The satellite frequencies are usually variable

B: When it is said that a satellite is operating in "U/V mode," this means that the satellite uplink is in the 70 cm band and the downlink is in the 2-meter band. In ham radio, modes are particular types of modulation, and modulations are the systems for adding information to a radio signal to be broadcast over the air.

T8B09. What causes spin fading of satellite signals?

A. Circular polarized noise interference radiated from the sun
B. Rotation of the satellite and its antennas
C. Doppler shift of the received signal
D. Interfering signals within the satellite uplink band

B: Spin fading with respect to satellite signals is caused by the rotation of the satellite and its antennas.

T8B10. What does the term LEO mean in reference to communication satellites?

A. Low Energy Orbit, which conserves battery power
B. Low Elevation Orbit, which appears close to the horizon from the earth station
C. Low Equilibrium Orbit, which has a slightly unstable period
D. Low Earth Orbit, which has a period of around 100 minutes

D: The initials LEO indicate that the satellite is in a Low Earth Orbit. This means that the path of the satellite is less than 1000 miles above the surface of the Earth.

T8B11. Who is permitted to receive telemetry from an amateur radio satellite?

A. Anyone
B. Only the satellite control operator
C. Only the control operator or a licensed radio amateur who has received the encryption key from the control operator
D. Only a licensed radio amateur who has received the encryption key from AMSAT

A: Anyone is permitted to receive telemetry from an amateur radio satellite.

T8B12. What is a way to determine whether your satellite uplink power into a linear transponder satellite is neither too low nor too high?

A. Check your signal strength report in the telemetry data
B. Listen for distortion on your downlink signal
C. Your signal strength on the downlink should be about the same as the beacon
D. Compare your signal to others on the downlink using an internet SDR receiver

C: Your signal strength on the downlink should be about the same as the beacon.

T8C – Operating Activities: Radio Direction Finding, Contests, Linking Over the Internet, Exchanging Grid Locators

T8C01. Which of the following methods is used to locate sources of noise interference or jamming?

A. Echolocation
B. Doppler radar
C. Radio direction finding
D. Phase locking

C: Radio direction finding is used to locate sources of noise interference or jamming.

T8C02. Which of these items would be useful for a hidden transmitter hunt?

A. Calibrated SWR meter
B. A directional antenna
C. A directional wattmeter
D. All these choices are correct

B: A directional antenna would be useful for a hidden transmitter hunt. Directional antennas send and receive signals in some directions more than others. Therefore, these antennas are able to ascertain from which direction a signal is being broadcast.

T8C03. What operating activity involves contacting as many stations as possible during a specified period?

A. Simulated emergency exercises
B. Net operations
C. Hidden transmitter hunts
D. Contesting

D: Contesting involves contacting as many stations as possible during a specified period of time.

T8C04. Which of the following is good practice when contacting another station in a contest?

A. Signing only the last two letters of your call if there are many other stations calling
B. Contacting the station twice to be sure that you are in his log
C. Sending only the minimum information needed for proper identification and the contest exchange
D. Adding "Please copy" before your exchange

C: Sending only the minimum information needed for proper identification and the contest exchange is good procedure when contacting another station in a radio contest.

T8C05. What is a grid locator?

A. A letter-number designator assigned to a geographic location
B. A letter-number designator assigned to an azimuth and elevation
C. An instrument for locating faults in power amplifiers
D. An instrument for radio direction finding

A: A grid locator is a letter-number designator assigned to a geographic location. The latitude and the longitude of the location are compressed into a brief string of characters, in alternating pairs of letters and numbers.

T8C06. How is over the air access to Internet Radio Linking Project (IRLP) nodes accomplished?

A. By obtaining a password that is sent via voice to the node
B. By using Dual-Tone Multi-Frequency (DTMF) signals
C. By entering the proper internet password
D. By using Continuous Tone-Coded Squelch System (CTCSS) tone codes

B: Internet Radio Linking Project (IRLP) nodes consist of a radio attached to an internet-connected computer running specialized software. It allows amateur radio users to communicate with other users anywhere in the world near another IRLP node. Users access the node using Dual-Tone Multi-Frequency (DTMF) signals.

T8C07. What is Voice Over Internet Protocol (VoIP)?

A. A set of rules specifying how to identify your station when linked over the internet to another station
B. A technique employed to "spot" DX stations via the internet
C. A technique for measuring the modulation quality of a transmitter using remote sites monitored via the internet
D. A method of delivering voice communications over the internet using digital techniques

D: Voice-over-Internet Protocol (VoIP) is a method of delivering voice communications over the Internet. It is an integral part of the Internet Radio Linking Project (IRLP).

T8C08. What is the Internet Radio Linking Project (IRLP)?

A. A technique to connect amateur radio systems, such as repeaters, via the internet
B. A system for providing access to websites via amateur radio
C. A system for informing amateurs in real time of the frequency of active DX stations
D. A technique for measuring signal strength of an amateur transmitter via the internet

A: The IRLP is a project that makes use of the Internet to connect amateur radio stations to one another from all over the world.

T8C09. Which of the following protocols enables an amateur station to transmit through a repeater without using a radio to initiate the transmission?

A. IRLP
B. D-STAR
C. DMR
D. EchoLink

D: EchoLink enables an amateur station to transmit through a repeater without using a radio to initiate the transmission.

T8C10. What is required before using the EchoLink system?

A. Complete the required EchoLink training
B. Purchase a license to use the EchoLink software
C. Register your call sign and provide proof of license
D. At least a General Class license

C: You must provide your call sign and proof of license since this type of use is a type of remote control of a station.

T8C11. What is an amateur radio station that connects other amateur stations to the internet?

A. A gateway
B. A repeater
C. A digipeater
D. A beacon

A: A gateway is an amateur radio station that is used to connect other amateur stations to the Internet.

T8D – Non-voice and digital communications: image signals and definition of NTSC, CW, packet radio, PSK, APRS, error detection and correction, amateur radio networking, DMR, WSJT modes, Broadband-Hamnet

T8D01. Which of the following is a digital communications mode?

A. Packet radio
B. IEEE 802.11
C. FT8
D. All these choices are correct

D: Packet, IEEE 802.11, and FT8 are all examples of digital communications methods.

T8D02. What is FT8?

A. A wideband FM voice mode
B. A digital mode capable of low signal-to-noise operation
C. An eight-channel multiplex mode for FM repeaters
D. A digital slow-scan TV mode with forward error correction and automatic color compensation

B: FT8 is a digital mode capable of low signal-to-noise operation.

T8D03. What kind of data can be transmitted by APRS?

A. GPS position data
B. Text messages
C. Weather data
D. All these choices are correct

D: GPS position data, text messages, and weather data can all be transmitted by APRS.

T8D04. What is meant by the term "NTSC?"

A. A digital transmission standard for encrypting data
B. A special mode for satellite uplink
C. An analog fast-scan color TV signal
D. A frame compression scheme for TV signals

C: The term NTSC indicates an analog fast scan color TV signal.

T8D05. Which of the following is an application of APRS?

A. Providing real-time tactical digital communications in conjunction with a map showing the locations of stations
B. Automatically showing the number of packets transmitted via PACTOR during a specific time interval
C. Providing voice over internet connection between repeaters
D. Providing information on the number of stations signed into a repeater

A: The Automatic Packet Reporting System (APRS) is a system designed to provide real-time tactical digital communications along with a map indicating the sources of the signals carrying the communication.

T8D06. What does the abbreviation "PSK" mean?

A. Pulse Shift Keying
B. Phase Shift Keying
C. Packet Sampled Keying
D. Power Sampled Keying

B: The abbreviation PSK means phase shift keying. Phase shift keying is a strategy for digital modes in the high frequency spectrum. In this system, the components of code are indicated with shifts in the timing relationship between a continuously transmitted tone and a reference tone. This enables the user to transmit a signal in very limited space.

T8D07. Which of the following describes DMR?

A. A technique for time-multiplexing two digital voice signals on a single 12.5 kHz repeater channel
B. An automatic position tracking mode for FM mobiles communicating through repeaters
C. An automatic computer logging technique for hands-off logging when communicating while operating a vehicle
D. A digital technique for transmitting on two repeater inputs simultaneously for automatic error correction

A: DMR is specifically a way to time-multiplex two signals on a single channel.

T8D08. Which of the following is included in packet radio transmissions?

A. A checksum that permits error detection
B. A header that contains the call sign of the station to which the information is being sent
C. Automatic repeat request in case of error
D. All these choices are correct

D: Packet transmissions may include a checksum which permits error detection, a header which contains the call sign of the station to which the information is being sent, and an automatic repeat request in case of error.

T8D09. What is CW?

A. A type of electromagnetic propagation
B. A digital mode used primarily on 2-meter FM
C. Error correction for digital transmission using code words
D. Another name for a Morse code transmission

D: CW is another name for a Morse code transmission.

T8D10. Which of the following operating activities is supported by digital mode software in the WSJT-X software suite?

A. Earth-Moon-Earth
B. Weak signal propagation beacons
C. Meteor scatter
D. All these choices are correct

D: Earth-Moon-Earth, weak signal propagation beacons, and meteor scatter are all well suited to the short transmissions of WSJT.

T8D11. What is the role of ARQ in a transmission system?

A. A special transmission format limited to video signals
B. A system used to encrypt command signals to an amateur radio satellite
C. An error correction method in which the receiving station detects errors and sends a request for retransmission
D. A method of compressing data using autonomous reiterative Q codes prior to final encoding

C: An Automatic Repeat Query (ARQ) transmission system is a system designed to catch errors in transmission. From the receiving station, it detects errors in the transmission and automatically sends a request to retransmit the information.

T8D12. Which of the following best describes an amateur radio mesh network?

A. An amateur-radio data network using commercial Wi-Fi equipment with modified firmware
B. A wide-bandwidth digital voice mode employing DMR protocols
C. An amateur-radio satellite communications network using modified commercial satellite TV hardware
D. An internet linking protocol allowing communication through repeaters around the world

A: An amateur radio mesh network is an amateur-radio based data network using commercial Wi-Fi equipment with modified firmware.

T9 – Antennas and Feed Lines

T9A – Antennas: Vertical and Horizontal Polarization, Concept of Antenna Gain, Definition and Types of Beam Antennas, Antenna Loading, Common Portable and Mobile Antennas, Relationships Between Resonant Length and Frequency, Dipole Pattern

T9A01. What is a beam antenna?

A. An antenna built from square aluminum beams
B. An omnidirectional antenna invented by Clarence Beam
C. An antenna that concentrates signals in one direction
D. An antenna that focuses the signal into two intense rays

C: A beam antenna concentrates signals in one direction. Beam antennas, also known as beamforming antennas, are sometimes composed of a simple system made of two wires. The most popular beam antenna is the Yagi, which consists of two conducting elements running parallel to each other and 1/10 of a wavelength apart.

T9A02. Which of the following describes a type of antenna loading?

A. Electrically lengthening by inserting inductors in radiating elements
B. Inserting a resistor in the radiating portion of the antenna to make it resonant
C. Installing a spring in the base of a mobile vertical antenna to make it more flexible
D. Strengthening the radiating elements of a beam antenna to better resist wind damage

A: Antenna loading is the term for placing an inductor in series with the antenna so that it reduces the resonant frequency of the antenna.

T9A03. How is the polarization of an antenna described?

A. By the shape of the driven element
B. By the orientation of the electric field
C. By the orientation of the magnetic field
D. By the direction of radiation

B: The polarization of an antenna is described by the orientation of the electric field.

T9A04. What is a disadvantage of a handheld radio transceiver's short flexible antenna compared to a full-sized quarter-wave antenna?

A. It has low efficiency
B. It transmits only circularly polarized signals
C. It is more susceptible to receiver desensitization
D. It only works on analog signals, not digital ones

A: One disadvantage of the antenna supplied with most handheld radio transceivers is it does not transmit or receive as effectively as a full-sized antenna.

T9A05. Which of the following increases the resonant frequency of a dipole antenna?

A. Lengthening it
B. Inserting coils in series with radiating wires
C. Shortening it
D. Adding capacitive loading to the ends of the radiating wires

C: To make a dipole antenna resonant on a higher frequency, shorten it.

T9A06. Which of the following types of antennas offers the greatest gain?

A. 5/8 wave vertical
B. Isotropic
C. J pole
D. Yagi

D: When compared to 5/8 wave vertical, isotropic, and J pole antennas, Yagi antennas offer the greatest gain.

T9A07. What is a potential drawback of using a handheld VHF transceiver inside a vehicle that lacks an externally mounted antenna?

A. Signal strength is reduced due to the shielding effect of the vehicle
B. The bandwidth of the antenna will decrease, increasing SWR
C. The SWR might decrease, decreasing the signal strength
D. The handheld will overheat due to reflected power in the vehicle

A: One good reason not to use a "rubber duck" antenna inside your car is that signals can be significantly weaker than when it is outside the vehicle.

T9A08. Why is a 19-inch-long vertical antenna often used on 2 meters?

A. It has high gain
B. It is a resonant half-wave
C. It is a resonant quarter-wave
D. It has low RF radiation exposure

C: A 19-inch-long vertical antenna is often used on 2 meters because it is a resonant quarter-wave.

T9A09. What is an advantage of a 5/8-wavelength whip antenna for VHF or UHF mobile service compared to a 1/4-wave antenna?

A. It has more gain
B. It radiates at a higher angle
C. It has lower SWR
D. It has lower impedance

A: A 5/8-wavelength whip antenna has an advantage over a 1/4-wave antenna because it has more gain.

T9A10. In which direction does a half-wave dipole antenna radiate the strongest signal?

A. Equally in all directions
B. Off the ends of the antenna
C. In the direction of the feed line
D. Broadside to the antenna

D: The radiation is strongest from a half-wave dipole antenna in free space broadside to the antenna.

T9A11. What is antenna gain?

A. The additional power that is added to the transmitter power
B. The additional power that is required in the antenna when transmitting on a higher frequency
C. The increase in signal strength in a specified direction compared to a reference antenna
D. The increase in impedance on receive or transmit compared to a reference antenna

C: The gain of an antenna is the increase in signal strength in a specified direction when compared to a reference antenna.

T9B – Feed lines: types, attenuation vs frequency, selecting; SWR concepts; Antenna tuners (couplers); RF Connectors: selecting, weather protection

T9B01. Which of the following connectors should be carefully taped for weather protection when used outdoors?

A. PL259
B. BNC
C. Type N
D. All these choices are correct

D: All these choices are correct because PL259, BNC, and Type N connectors should all be carefully taped for weather protection when used outdoors.

T9B02. What is the most common impedance of coaxial cables used in amateur radio?

A. 8 ohms
B. 50 ohms
C. 600 ohms
D. 12 ohms

B: The impedance of the most commonly used coaxial cable in typical amateur radio installations is 50 ohms. Impedance is the measure of the total opposition to current flow in an alternating current circuit, composed of reactance and ohmic resistance.

T9B03. Why is coaxial cable the most common feed line for amateur radio antenna systems?

A. It is easy to use and requires few special installation considerations
B. It has less loss than any other type of feed line
C. It can handle more power than any other type of feed line
D. It is less expensive than any other type of feed line

A: Coaxial cable is used more often than any other feed lines for amateur radio antenna systems because it is easy to use and requires few special installation considerations.

T9B04. What is the major function of an antenna tuner (antenna coupler)?

A. It matches the antenna system impedance to the transceiver's output impedance
B. It helps a receiver automatically tune in weak stations
C. It allows an antenna to be used on both transmit and receive
D. It automatically selects the proper antenna for the frequency band being used

A: An antenna tuner matches the antenna system impedance to the transceiver's output impedance.

T9B05. What happens as the frequency of a signal in coaxial cable is increased?

A. The characteristic impedance decreases
B. The loss decreases
C. The characteristic impedance increases
D. The loss increases

D: As the frequency of a signal passing through coaxial cable is increased, the loss generally increases as well.

T9B06. Which of the following connector types is most suitable as an RF connector for frequencies above 400 MHz?

A. PL-259
B. Type N
C. RS-213
D. DB-25

B: A Type N connector is most suitable for frequencies above 400 MHz. Type N connectors are threaded RF connectors used to join coaxial cables. These cables are able to carry microwave frequency signals.

T9B07. Which of the following is true of PL-259 type coax connectors?

A. They are preferred for microwave operation
B. They are watertight
C. They are commonly used at HF and VHF frequencies
D. They are a bayonet-type connector

C: PL–259 type coax connectors are commonly used at HF and VHF frequencies.

T9B08. Which of the following is a source of loss in coaxial feed line?

A. Water intrusion into coaxial connectors
B. High SWR
C. Multiple connectors in the line
D. All these choices are correct

D: Water intrusion into coaxial connectors, high SWR, and multiple connectors in the line are all sources of loss in a coaxial feed line.

T9B09. What can cause erratic changes in SWR?

A. Local thunderstorm
B. Loose connection in the antenna or feed line
C. Over-modulation
D. Overload from a strong local station

B: A loose connection in an antenna or a feed line might cause erratic changes in SWR readings.

T9B10. What is the electrical difference between RG-58 and RG-213 coaxial cable?

A. There is no significant difference between the two types
B. RG-58 cable has two shields
C. RG-213 cable has less loss at a given frequency
D. RG-58 cable can handle higher power levels

C: An RG-213 cable has less loss at a given frequency than an RG-58 cable.

T9B11. Which of the following types of feed line has the lowest loss?

A. 50-ohm flexible coax
B. Multi-conductor unbalanced cable
C. Air-insulated hardline
D. 75-ohm flexible coax

C: Air-insulated hardline is the type of feed line that has the lowest loss.

T9B12. What is standing wave ratio (SWR)?

A. A measure of how well a load is matched to a transmission line
B. The ratio of amplifier power output to input
C. The transmitter efficiency ratio
D. An indication of the quality of your station's ground connection

A: In general terms, standing wave ratio (SWR) is a measure of how well a load is matched to a transmission line. This is a measure of the impedance match between the antenna and the feedline. It is the ratio of the maximum voltage to the minimum voltage along the feedline. If a transmitter is being used, standing wave ratio is the impedance match between the feedline from the transmitter and the antenna system. If the antenna is a purely resistive load, the standing wave ratio is the ratio of antenna impedance to feedline impedance.

T0 –Safety

T0A – Power circuits and hazards: hazardous voltages, fuses and circuit breakers, grounding, electrical code compliance; Lightning protection; Battery safety

T0A01. Which of the following is a safety hazard of a 12-volt storage battery that lacks internal protection circuitry?

A. Touching both terminals with your hands can cause electrical shock
B. Shorting the terminals can cause burns, fire, or an explosion
C. RF emissions from a nearby transmitter can cause the electrolyte to emit poison gas
D. All these choices are correct

B: In order for electrical shock from a 12-volt battery to be hazardous, the circuit would typically need to be connected on or near one's chest. The primary hazard associated with a 12-volt battery is shorting the terminals with low resistance materials such as metal wires or rods, potentially resulting in burns, fires, or explosions.

T0A02. What health hazard is posed by electrical current flowing through the body?

A. It may cause injury by heating body tissue
B. It may disrupt the electrical functions of cells
C. It may cause involuntary muscle contractions
D. All these choices are correct

D: Electric shock may result in any of the following: cell damage, electrical function disruption in cells, tissue heating, muscle contraction, or burns.

T0A03. In the United States, what circuit does black wire insulation indicate in a three-wire 120 V AC cable?

A. Neutral
B. Hot
C. Equipment ground
D. Negative

B: In a three-wire electrical AC plug, the black wire is hot.

T0A04. What is the purpose of a fuse in an electrical circuit?

A. To prevent power supply ripple from damaging a component
B. To remove power in case of an overload
C. To limit current and prevent shocks
D. All these choices are correct

B: The purpose of a fuse in an electrical circuit is to interrupt power in case of overload.

T0A05. Why should a 5-ampere fuse never be replaced with a 20-ampere fuse?

A. The larger fuse would be likely to blow because it is rated for higher current
B. The power supply ripple would greatly increase
C. Excessive current could cause a fire
D. Voltage drop in the higher current fuse could result in excessively low voltage to the device

C: It is unwise to install a 20-ampere fuse in the place of a 5-ampere fuse because the excessive current could cause a fire.

T0A06. What is a good way to guard against electrical shock at your station?

A. Use three-wire cords and plugs for all AC powered equipment
B. Connect all AC powered station equipment to a common safety ground
C. Ensure all capacitors used for high-voltage DC are fully discharged before working inside equipment
D. All these choices are correct

D: There are several good ways to guard against electrical shock at your station. Use three-wire cords and plugs for all AC-powered equipment. Connect all AC-powered station equipment to a common safety ground. Also, ensure that capacitors used for high-voltage DC are fully discharged before working inside equipment.

T0A07. Where should a lightning arrester be installed in a coaxial feed line?

A. At the output connector of a transceiver
B. At the antenna feed point
C. At the AC power service panel
D. On a grounded panel near where feed lines enter the building

D: When installing a lightning arrester in a coaxial cable feed line, it should be near where the feed lines enter the building and grounded to a common plate which is in turn connected to an external ground.

T0A08. Where should a fuse or circuit breaker be installed in a 120V AC power circuit?

A. In series with the hot conductor only
B. In series with the hot and neutral conductors
C. In parallel with the hot conductor only
D. In parallel with the hot and neutral conductors

A: Home-built equipment that is powered with 120V AC power should always include a fuse or circuit breaker in series with the hot power line to reduce the risk of electrocution hazard.

T0A09. What should be done to all external ground rods or earth connections?

A. Waterproof them with silicone caulk or electrical tape
B. Keep them as far apart as possible
C. Bond them together with heavy wire or conductive strap
D. Tune them for resonance on the lowest frequency of operation

C: The best practice for ensuring safety regarding ground rods and earth connections is to bond them together. This can be done with heavy gauge wire or a large conductive strap.

T0A10. What hazard exists when rapidly charging or discharging an unprotected battery?

A. Overheating or out-gassing
B. Excess output ripple
C. Electric shock
D. Overvoltage

A: If a battery is charged or discharged too quickly, the battery could overheat and give off flammable gas, or explode.

T0A11. What hazard exists in a power supply immediately after turning it off?

A. Circulating currents in the dc filter
B. Leakage flux in the power transformer
C. Voltage transients from kickback diodes
D. Charge stored in filter capacitors

D: When a power supply is turned off and disconnected, a person still might receive an electric shock from the stored charge in large capacitors.

T0A12. Which of the following precautions should be taken when measuring high voltages with a voltmeter?

A. Ensure that the voltmeter has very low impedance
B. Ensure that the voltmeter and its leads are rated for use at the voltages being measured
C. Ensure that the circuit is grounded through the voltmeter
D. Ensure that the voltmeter is set to the correct frequency

B: When dealing with high voltages, safety should always be your foremost concern. If you need to measure the voltage on a high voltage circuit, always make sure that the voltmeter you are using and the attached leads are rated for use at the voltages you will be measuring.

T0B – Antenna Safety: Tower Safety and Grounding, Installing Antennas, Antenna Supports

T0B01. Which of the following is good practice when installing ground wires on a tower for lightning protection?

A. Put a drip loop in the ground connection to prevent water damage to the ground system
B. Make sure all ground wire bends are right angles
C. Ensure that connections are short and direct
D. All these choices are correct

C: When installing ground wires on a tower for lightning protection it is always good practice to ensure that connections are short and direct.

T0B02. What is required when climbing an antenna tower?

A. Have sufficient training on safe tower climbing techniques
B. Use appropriate tie-off to the tower at all times
C. Always wear an approved climbing harness
D. All these choices are correct

D: When climbing an antenna tower it is essential that you have sufficient training on safe tower climbing techniques, use appropriate tie-off to the tower at all times, and wear an approved climbing harness.

T0B03. Under what circumstances is it safe to climb a tower without a helper or observer?

A. When no electrical work is being performed
B. When no mechanical work is being performed
C. When the work being done is not more than 20 feet above the ground
D. Never

D: It is never safe to climb a tower without a helper or observer.

T0B04. Which of the following is an important safety precaution to observe when putting up an antenna tower?

A. Wear a ground strap connected to your wrist at all times
B. Insulate the base of the tower to avoid lightning strikes
C. Look for and stay clear of any overhead electrical wires
D. All these choices are correct

C: When putting up an antenna tower, it is important to look for and stay clear of any overhead electric wires.

T0B05. What is the purpose of a safety wire through a turnbuckle used to tension guy lines?

A. Secure the guy line if the turnbuckle breaks
B. Prevent loosening of the turnbuckle from vibration
C. Provide a ground path for lightning strikes
D. Provide an ability to measure for proper tensioning

B: The safety wire prevents the turnbuckle from turning. This will mediate any potential loosening of the turnbuckle due to vibration.

T0B06. What is the minimum safe distance from a power line to allow when installing an antenna?

A. Add the height of the antenna to the height of the power line and multiply by a factor of 1.5
B. The height of the power line above ground
C. 1/2 wavelength at the operating frequency
D. Enough so that if the antenna falls, no part of it can come within 10 feet of the power wires

D: The minimum safe distance from a power line when installing an antenna is such that, if the antenna falls unexpectedly, no part of it can come closer than 10 feet to the power wires.

T0B07. Which of the following is an important safety rule to remember when using a crank-up tower?

A. This type of tower must never be painted
B. This type of tower must never be grounded
C. This type of tower must not be climbed unless it is retracted, or mechanical safety locking devices have been installed
D. All these choices are correct

C: An important safety rule to remember when using a crank-up tower is that this type of tower must never be climbed unless it is in the fully retracted position. Crank-up towers are kept stable by a concrete foundation rather than by guy wires.

T0B08. Which is a proper grounding method for a tower?

A. A single four-foot ground rod, driven into the ground no more than 12 inches from the base
B. A ferrite-core RF choke connected between the tower and ground
C. A connection between the tower base and a cold-water pipe
D. Separate eight-foot ground rods for each tower leg, bonded to the tower and each other

D: Separate eight-foot long ground rods for each tower leg, bonded to the tower and each other, are considered to be a proper grounding method for a tower.

T0B09. Why should you avoid attaching an antenna to a utility pole?

A. The antenna will not work properly because of induced voltages
B. The antenna may unbalance the power transformer, causing power fluctuations
C. The antenna could contact high-voltage power lines
D. All these choices are correct

C: You should avoid attaching an antenna to a utility pole because the antenna could contact high-voltage power wires.

T0B10. Which of the following is true when installing grounding conductors used for lightning protection?

A. Use only non-insulated wire
B. Wires must be carefully routed with precise right-angle bends
C. Sharp bends must be avoided
D. Common grounds must be avoided

C: Sharp bends must be avoided in the grounding conductors used for lightning protection.

T0B11. Which of the following establishes grounding requirements for an amateur radio tower or antenna?

A. FCC Part 97 rules
B. Local electrical codes
C. FAA tower lighting regulations
D. UL recommended practices

B: Local electrical codes establish grounding requirements for an amateur radio tower or antenna.

T0C – RF HAZARDS: RADIATION EXPOSURE, PROXIMITY TO ANTENNAS, RECOGNIZED SAFE POWER LEVELS, RADIATION TYPES, DUTY CYCLE

T0C01. What type of radiation are radio signals?

A. Gamma radiation
B. Ionizing radiation
C. Alpha radiation
D. Non-ionizing radiation

D: Radio signals are nonionizing radiation. That is, they are electromagnetic radiation without enough energy to knock electrons off of atoms or molecules, which would create an ion.

T0C02. Which of the following bands has the lowest maximum permissible exposure for RF safety?

A. 3.5 MHz
B. 50 MHz
C. 440 MHz
D. 1296 MHz

B: 50 MHz has a lower maximum permissible exposure limit than 3.5 MHz, 440 MHz, or 1296 MHz.

T0C03. How does the allowable power density for RF safety change if duty cycle changes from 100 percent to 50 percent?

A. It increases by a factor of 3
B. It decreases by 50 percent
C. It increases by a factor of 2
D. There is no adjustment allowed for lower duty cycle

C: When duty cycle changes from 100 percent to 50 percent, the allowable power density for RF safety increases by a factor of 2.

T0C04. What factors affect the RF exposure of people near an amateur station antenna?

A. Frequency and power level of the RF field
B. Distance from the antenna to a person
C. Radiation pattern of the antenna
D. All these choices are correct

D: Several factors affect the RF exposure of people near an amateur station antenna. These factors include the frequency and power level of the RF field, the distance from the antenna to a person, and the radiation pattern of the antenna.

T0C05. Why do exposure limits vary with frequency?

A. Lower frequency RF fields have more energy than higher frequency fields
B. Lower frequency RF fields do not penetrate the human body
C. Higher frequency RF fields are transient in nature
D. The human body absorbs more RF energy at some frequencies than at others

D: Exposure limits vary with frequency because the human body absorbs more RF energy at some frequencies than at others.

T0C06. Which of the following is an acceptable method to determine whether your station complies with FCC RF exposure regulations?

A. By calculation based on FCC OET Bulletin 65
B. By calculation based on computer modeling
C. By measurement of field strength using calibrated equipment
D. All these choices are correct

D: It is acceptable to determine that your station complies with FCC RF exposure regulations by calculation based on FCC OET Bulletin 65, by calculation based on computer modeling, or by measurement of field strength using calibrated equipment.

T0C07. What hazard is created by touching an antenna during a transmission?

A. Electrocution
B. RF burn to skin
C. Exposure to ionizing radiation
D. All these choices are correct

B: If a person accidentally touched your antenna while you were transmitting, they might receive a painful RF burn.

T0C08. Which of the following actions can reduce exposure to RF radiation?

A. Relocate antennas
B. Relocate the transmitter
C. Increase the duty cycle
D. All these choices are correct

A: Amateur operators might relocate antennas to prevent exposure to RF radiation in excess of FCC-supplied limits.

T0C09. How can you make sure your station stays in compliance with RF safety regulations?

A. By informing the FCC of any changes made in your station
B. By re-evaluating the station whenever an item in the transmitter or antenna system is changed
C. By making sure your antennas have low SWR
D. By using only Underwriter Laboratories approved transmitting equipment

B: A station may ensure compliance by re-evaluating the station whenever an item of equipment is changed.

T0C10. Why is duty cycle one of the factors used to determine safe RF radiation exposure levels?

A. It affects the average exposure to radiation
B. It affects the peak exposure to radiation
C. It takes into account the antenna feed line loss
D. It takes into account the thermal effects of the final amplifier

A: Duty cycle is one of the factors used to determine safe RF radiation exposure levels because it affects the average exposure of people to radiation. If the duty cycle is lower, then there is less RF radiation exposure with no change in peak envelope power output.

T0C11. What is the definition of duty cycle during the averaging time for RF exposure?

A. The difference between the lowest and highest power output of a transmitter
B. The difference between the PEP and the average power output of a transmitter
C. The percentage of time that a transmitter is transmitting
D. The percentage of time that a transmitter is not transmitting

C: In the context of determining RF exposure, the duty cycle is the percentage of the time that the transmitter is actually transmitting.

T0C12. How does RF radiation differ from ionizing radiation (radioactivity)?

A. RF radiation does not have sufficient energy to cause chemical changes in cells and damage DNA
B. RF radiation can only be detected with an RF dosimeter
C. RF radiation is limited in range to a few feet
D. RF radiation is perfectly safe

A: Unlike most ionizing radiation, RF radiation does not carry sufficient energy to cause genetic damage to living creatures.

T0C13. Who is responsible for ensuring that no person is exposed to RF energy above the FCC exposure limits?

A. The FCC
B. The station licensee
C. Anyone who is near an antenna
D. The local zoning board

B: The station licensee is responsible for ensuring that no person is exposed to RF energy above the FCC exposure limits.

HAM Radio Technician Practice Tests

Test #1 Practice Questions

1. Which of the following is part of the Basis and Purpose of the Amateur Radio Service?

A. Providing personal radio communications for as many citizens as possible
B. Providing communications for international contesting
C. Advancing skills in the technical and communication phases of the radio art
D. All these choices are correct

2. Which of the following frequency ranges are available for phone operation by Technician licensees?

A. 28.050 MHz to 28.150 MHz
B. 28.100 MHz to 28.300 MHz
C. 28.300 MHz to 28.500 MHz
D. 28.500 MHz to 28.600 MHz

3. For which classes of amateur radio licenses does the FCC currently issue new licenses?

A. Novice, Technician, General, Amateur Extra
B. Technician, Technician Plus, General, Amateur Extra
C. Novice, Technician Plus, General, Advanced
D. Technician, General, Amateur Extra

4. With which countries are FCC-licensed amateur radio stations prohibited from exchanging communications?

A. Any country whose administration has notified the International Telecommunication Union (ITU) that it objects to such communications
B. Any country whose administration has notified the American Radio Relay League (ARRL) that it objects to such communications
C. Any country banned from such communications by the International Amateur Radio Union (IARU)
D. Any country banned from making such communications by the American Radio Relay League (ARRL)

5. When may an amateur station transmit without a control operator?

A. When using automatic control, such as in the case of a repeater
B. When the station licensee is away and another licensed amateur is using the station
C. When the transmitting station is an auxiliary station
D. Never

6. When must the station licensee make the station and the station records available for inspection?

A. At any time ten days after notification by the FCC of such an inspection
B. At any time upon request by an FCC representative
C. At any time after a written notification by the FCC of such inspection
D. Only when presented with a valid warrant by an FCC official or government agent

7. What is a common repeater frequency offset in the 2-meter band?

A. Plus or minus 5 MHz
B. Plus or minus 600 kHz
C. Plus or minus 500 kHz
D. Plus or minus 1 MHz

8. What is the purpose of the reverse function on a VHF/UHF transceiver?

A. To reduce power output
B. To increase power output
C. To listen on a repeater's input frequency
D. To listen on a repeater's output frequency

9. When do the FCC Part 97 Amateur Radio Service rules NOT apply to the operation of an amateur station?

A. When operating under RACES rules
B. When operating under FEMA rules
C. When operating under ARES rules
D. FCC rules always apply

10. Why do VHF signal strengths sometimes vary greatly when the antenna is moved only a few feet?

A. The signal path encounters different concentrations of water vapor
B. VHF ionospheric propagation is very sensitive to path length
C. Multipath propagation cancels or reinforces signals
D. The Doppler effect causes slight frequency shifts which result in changes in signal strength

11. What is the relationship between the electric and magnetic fields of an electromagnetic wave?

A. They travel at different speeds
B. They are in parallel
C. They revolve in opposite directions
D. They are at right angles

12. Why are simplex UHF signals rarely heard beyond their radio horizon?

A. They are too weak to go very far
B. FCC regulations prohibit them from going more than 50 miles
C. UHF signals are usually not propagated by the ionosphere
D. UHF signals are absorbed by the ionospheric D region

13. Which of the following is an appropriate power supply rating for a typical 50-watt output mobile FM transceiver?

A. 24.0 volts at 4 amperes
B. 13.8 volts at 4 amperes
C. 24.0 volts at 12 amperes
D. 13.8 volts at 12 amperes

14. What is the effect of excessive microphone gain on SSB transmissions?

A. Frequency instability
B. Distorted transmitted audio
C. Increased SWR
D. Sideband inversion

15. Electrical current is measured in which of the following units?

A. Volts
B. Watts
C. Ohms
D. Amperes

16. How many milliamperes is 1.5 amperes?

A. 0.0000015 milliamperes
B. 0.0015 milliamperes
C. 1500 milliamperes
D. 1,500,000 milliamperes

17. What describes the ability to store energy in an electric field?

A. Inductance
B. Resistance
C. Frequency
D. Capacitance

18. What formula is used to calculate current in a circuit?

A. I = E × R
B. I = E / R
C. I = E2 × R
D. I = E2 / R

19. What electrical component opposes the flow of current in a DC circuit?

A. Inductor
B. Resistor
C. Inverter
D. Transformer

20. Which is true about forward voltage drop in a diode?

A. It is lower in some diode types than in others
B. It is proportional to peak inverse voltage
C. It indicates that the diode is defective
D. It has no impact on the voltage delivered to the load

21. What is an electrical diagram using standard component symbols called?

A. Connection chart
B. Instrumentation system
C. Schematic
D. Flow chart

22. Which of the following devices or circuits changes an alternating current into a varying direct current signal?

A. Transformer
B. Rectifier
C. Amplifier
D. Reflector

23. Which term describes the ability of a receiver to detect the presence of a signal?

A. RF gain
B. Sensitivity
C. Selectivity
D. Total Harmonic Distortion

24. What can you do if you are told your FM handheld or mobile transceiver is over-deviating?

A. Talk louder into the microphone
B. Let the transceiver cool off
C. Change to a higher power level
D. Talk farther away from the microphone

25. What is the primary purpose of a dummy load?

A. To prevent transmitting signals over the air when making tests
B. To prevent over-modulation of a transmitter
C. To improve the efficiency of an antenna
D. To improve the signal-to-noise ratio of a receiver

26. Which instrument would you use to measure electric potential?

A. An ammeter
B. A voltmeter
C. A potentiometer
D. An ohmmeter

27. Which of the following is a form of amplitude modulation?

A. Spread spectrum
B. Packet radio
C. Single sideband
D. Phase shift keying (PSK)

28. What telemetry information is typically transmitted by satellite beacons?

A. The signal strength of received signals
B. Time of day accurate to plus or minus 1/10 second
C. Health and status of the satellite
D. All these choices are correct

29. Which of the following methods is used to locate sources of noise interference or jamming?

A. Echolocation
B. Doppler radar
C. Radio direction finding
D. Phase locking

30. Which of the following is a digital communications mode?

A. Packet radio
B. IEEE 802.11
C. FT8
D. All these choices are correct

31. What is a beam antenna?

A. An antenna built from square aluminum beams
B. An omnidirectional antenna invented by Clarence Beam
C. An antenna that concentrates signals in one direction
D. An antenna that focuses the signal into two intense rays

32. Which of the following connectors should be carefully taped for weather protection when used outdoors?

A. PL259
B. BNC
C. Type N
D. All these choices are correct

33. Which of the following is a safety hazard of a 12-volt storage battery that lacks internal protection circuitry?

A. Touching both terminals with your hands can cause electrical shock
B. Shorting the terminals can cause burns, fire, or an explosion
C. RF emissions from a nearby transmitter can cause the electrolyte to emit poison gas
D. All these choices are correct

34. Which of the following is good practice when installing ground wires on a tower for lightning protection?

A. Put a drip loop in the ground connection to prevent water damage to the ground system
B. Make sure all ground wire bends are right angles
C. Ensure that connections are short and direct
D. All these choices are correct

35. What type of radiation are radio signals?

A. Gamma radiation
B. Ionizing radiation
C. Alpha radiation
D. Non-ionizing radiation

Test #1 Answer Key and Explanations

1. C: As stated in the FCC rules, advancing skills in the technical and communication phases of the radio art is a purpose of the Amateur Radio Service

2. C: Frequencies in the 28.300 MHz to 28.500 MHz range are available for phone operation by Technician licensees. [97.301(e)]

3. D: The classes of amateur operator license grants are: Technician, General, and Amateur Extra. The person named in the operator license grant is authorized to be the control operator of an amateur station with the privileges authorized to the operator class specified on the license grant. [97.9(a), 97.17(a)]

4. A: FCC-licensed amateur stations are prohibited from exchanging communications with any country whose administration has notified the ITU that it objects to such communications. [97.111(a)(1)]

5. D: When transmitting, each amateur station must have a control operator. [97.7(a)]

6. B: The station licensee must make the station and the station records available for inspection upon request by an FCC representative. [97.103(c)]

7. B: The most common repeater frequency offset in the 2-meter band is plus or minus 600 kHz.

8. C: A VHF/UHF transceiver's "reverse" function is used to listen on a repeater's input frequency.

9. D: FCC rules always apply to the operation of an amateur station. [97.103(a)]

10. C: If a signal is experiencing multipath propagation, it can easily cancel itself out or reinforce itself over a distance of a few feet.

11. D: The electric and magnetic fields of electromagnetic waves are at right angles (perpendicular) to each other.

12. C: Simplex (that is, not by means of a repeater) UHF signals are rarely heard from stations outside the local coverage area because UHF signals are usually not reflected by the ionosphere.

13. D: For a typical 50 watt output mobile FM transceiver, 13.8 volts at 12 amperes would be an appropriate power supply rating.

14. B: If a transmitter is operated with the microphone gain set too high, the output signal might become distorted.

15. D: Electrical current is measured in amperes, usually referred to as amps. Current is the amount of electrons flowing through a circuit.

16. C: 1.5 amperes is equivalent to 1500 mA. The prefix milli- means thousandth, so a milliampere is a thousandth of an ampere. Therefore, 1.5 amperes can be converted to milliamperes by multiplying by a thousand: 1.5 x 1000 = 1500.

17. D: The ability to store energy in an electric field is called capacitance. Capacitance can also be defined as the ratio of the electric charge that is shifted from one conductor to another to the resulting potential difference between them.

18. B: The formula for calculating current in a circuit is current equals voltage divided by resistance.

19. B: The electrical component used to oppose the flow of current in a DC circuit is a resistor. For ham radio technicians, the most commonly required resistors are quarter- and half-watt fixed value resistors and 100, 500, 1k, 5k, 10k, and 100k ohm variable resistors.

20. A: Forward voltage drop is lower in some diode types than in others.

21. C: An electrical diagram with standard wiring component symbols is called a schematic.

22. B: A rectifier changes an alternating current into a varying direct current signal. Specifically, it is a switch that blocks the current in one direction but permits it in the other.

23. B: The ability of a receiver to detect the presence of a signal is measured in terms of its sensitivity.

24. D: If you are told your FM handheld or mobile transceiver is over deviating, you can talk farther away from the microphone.

25. A: The primary purpose of a dummy load is to prevent the radiation of signals when making tests. The dummy load, also known as a dummy antenna, is a station accessory that enables the user to test and finetune equipment without actually broadcasting a signal.

26. B: A voltmeter is used to measure electric potential or electromotive force. It accomplishes this by identifying the electric potential difference between two conductors.

27. C: Single sideband is a form of amplitude modulation. Indeed, single sideband is the most common mode of voice transmission on the high-frequency bands. Unlike an a.m. transmission, in which two identical versions of the message are broadcast, an SSB transmission only broadcasts one. For this reason, single sideband communications are more efficient and are better at conserving space on the radio spectrum.

28. C: Health and status are the most common telemetry datapoints that are transmitted by satellite beacons.

29. C: Radio direction finding is used to locate sources of noise interference or jamming.

30. D: Packet, IEEE 802.11, and FT8 are all examples of digital communications methods.

31. C: A beam antenna concentrates signals in one direction. Beam antennas, also known as beamforming antennas, are sometimes composed of a simple system made of two wires. The most popular beam antenna is the Yagi, which consists of two conducting elements running parallel to each other and 1/10 of a wavelength apart.

32. D: All these choices are correct because PL259, BNC, and Type N connectors should all be carefully taped for weather protection when used outdoors.

33. B: In order for electrical shock from a 12-volt battery to be hazardous, the circuit would typically need to be connected on or near one's chest. The primary hazard associated with a 12-volt battery is shorting the terminals with low resistance materials such as metal wires or rods, potentially resulting in burns, fires, or explosions.

34. C: When installing ground wires on a tower for lightning protection it is always good practice to ensure that connections are short and direct.

35. D: Radio signals are nonionizing radiation. That is, they are electromagnetic radiation without enough energy to knock electrons off of atoms or molecules, which would create an ion.

Test #2 Practice Questions

1. Which agency regulates and enforces the rules for the Amateur Radio Service in the United States?

A. ARRL
B. Homeland Security
C. The FCC
D. All these choices are correct

2. Which of the following U.S. amateur radio operators are allowed to contact the International Space Station (ISS) on VHF bands?

A. Only amateurs with a General class or higher license
B. Any amateur with a Technician class or higher license
C. Only amateurs with a General class or higher license, and NASA approval
D. Any amateurs with a Technician class or higher license, and NASA approval

3. Who may select a desired call sign under the vanity call sign rules?

A. Only a licensed amateur with a General or Amateur Extra Class license
B. Only a licensed amateur with an Amateur Extra Class license
C. Only a licensed amateur who has been licensed continuously for more than 10 years
D. Any licensed amateur

4. Under which of the following circumstances are one-way transmissions by an amateur station prohibited?

A. Announcements of upcoming ham radio operating events
B. Broadcasting
C. International Morse Code Practice
D. Telecommand or transmissions of telemetry

5. Who may be the control operator of a station communicating through an amateur satellite or space station?

A. Only an Amateur Extra Class operator
B. A General class or higher licensee with a satellite operator certification
C. Only an Amateur Extra Class operator who is also an AMSAT member
D. Any amateur allowed to transmit on the satellite uplink frequency

6. How often must you identify with your FCC-assigned call sign when using tactical call signs such as "Race Headquarters"?

A. Never, the tactical call is sufficient
B. Once every hour
C. At least every 10 minutes during and at the end of a communication
D. At the end of every transmission

7. What is the national calling frequency for FM simplex operations in the 2-meter band?

A. 146.520 MHz
B. 145.000 MHz
C. 432.100 MHz
D. 446.000 MHz

8. What term describes the use of a sub-audible tone transmitted along with normal voice audio to open the squelch of a receiver?

A. Carrier squelch
B. Tone burst
C. DTMF
D. CTCSS

9. Which of the following are typical duties of a Net Control Station?

A. Choose the regular net meeting time and frequency
B. Ensure that all stations checking into the net are properly licensed for operation on the net frequency
C. Call the net to order and direct communications between stations checking in
D. All these choices are correct

10. How does vegetation affect UHF and microwave signals?

A. Causes knife-edge diffraction, distorting voice peaks
B. Absorbs signals, leading to poor reception of weak signals
C. Amplifies signals, improving reception of weak signals
D. Has no effect

11. What property of a radio wave defines its polarization?

A. The orientation of the electric field
B. The orientation of the magnetic field
C. The ratio of the energy in the magnetic field to the energy in the electric field
D. The ratio of the velocity to the wavelength

12. What is a characteristic of HF communication compared with communications on VHF and higher frequencies?

A. HF antennas are generally smaller
B. HF accommodates wider bandwidth signals
C. Long-distance ionospheric propagation is far more common on HF
D. There is less atmospheric interference (static) on HF

13. Which of the following should be considered when selecting an accessory SWR meter?

A. The frequency and power level at which the measurements will be made
B. The distance that the meter will be located from the antenna
C. The maximum SWR anticipated on the transmission line
D. The ability of the meter to compensate for a poor transmission line match to the antenna

14. Which of the following can be used to enter a transceiver's operating frequency?

A. The keypad or VFO knob
B. The CTCSS or DTMF encoder
C. The Automatic Frequency Control
D. All these choices are correct

15. Electrical power is measured in which of the following units?

A. Volts
B. Watts
C. Watt-hours
D. Amperes

16. Which is equal to 1,500,000 hertz?

A. 1500 kHz
B. 1500 MHz
C. 15 GHz
D. 150 kHz

17. What is the unit of capacitance?

A. Farad
B. Ohm
C. Volt
D. Henry

18. What formula is used to calculate voltage in a circuit?

A. $E = I \times R$
B. $E = I / R$
C. $E = I2 \times R$
D. $E = I2 / R$

19. What type of component is often used as an adjustable volume control?

A. Fixed resistor
B. Power resistor
C. Potentiometer
D. Transformer

20. What electronic component allows current to flow in only one direction?

A. Resistor
B. Fuse
C. Diode
D. Driven element

21. What is component 1 in the figure below?

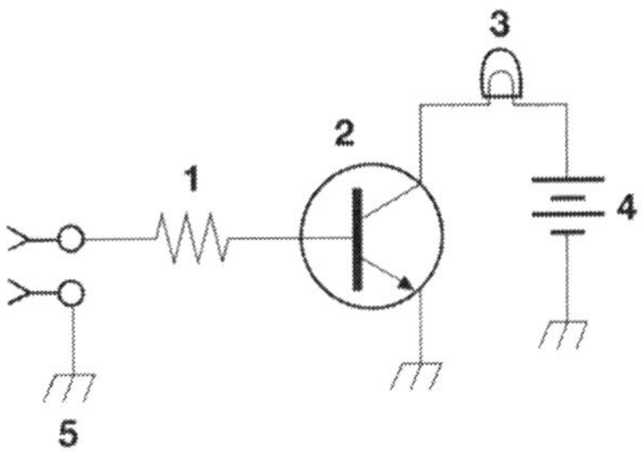

A. Resistor
B. Transistor
C. Battery
D. Connector

22. What is a relay?

A. An electrically-controlled switch
B. A current-controlled amplifier
C. An inverting amplifier
D. A pass transistor

23. What is a transceiver?

A. A device that combines a receiver and transmitter
B. A device for matching feed line impedance to 50 ohms
C. A device for automatically sending and decoding Morse code
D. A device for converting receiver and transmitter frequencies to another band

24. What would cause a broadcast AM or FM radio to receive an amateur radio transmission unintentionally?

A. The receiver is unable to reject strong signals outside the AM or FM band
B. The microphone gain of the transmitter is turned up too high
C. The audio amplifier of the transmitter is overloaded
D. The deviation of an FM transmitter is set too low

25. Which of the following is used to determine if an antenna is resonant at the desired operating frequency?

A. A VTVM
B. An antenna analyzer
C. A Q meter
D. A frequency counter

26. How is a voltmeter connected to a component to measure applied voltage?

A. In series
B. In parallel
C. In quadrature
D. In phase

27. What type of modulation is commonly used for VHF packet radio transmissions?

A. FM or PM
B. SSB
C. AM
D. PSK

28. What is the impact of using excessive effective radiated power on a satellite uplink?

A. Possibility of commanding the satellite to an improper mode
B. Blocking access by other users
C. Overloading the satellite batteries
D. Possibility of rebooting the satellite control computer

29. Which of these items would be useful for a hidden transmitter hunt?

A. Calibrated SWR meter
B. A directional antenna
C. A directional wattmeter
D. All these choices are correct

30. What is FT8?

A. A wideband FM voice mode
B. A digital mode capable of low signal-to-noise operation
C. An eight-channel multiplex mode for FM repeaters
D. A digital slow-scan TV mode with forward error correction and automatic color compensation

31. Which of the following describes a type of antenna loading?

A. Electrically lengthening by inserting inductors in radiating elements
B. Inserting a resistor in the radiating portion of the antenna to make it resonant
C. Installing a spring in the base of a mobile vertical antenna to make it more flexible
D. Strengthening the radiating elements of a beam antenna to better resist wind damage

32. What is the most common impedance of coaxial cables used in amateur radio?

A. 8 ohms
B. 50 ohms
C. 600 ohms
D. 12 ohms

33. What health hazard is posed by electrical current flowing through the body?

A. It may cause injury by heating body tissue
B. It may disrupt the electrical functions of cells
C. It may cause involuntary muscle contractions
D. All these choices are correct

34. What is required when climbing an antenna tower?

A. Have sufficient training on safe tower climbing techniques
B. Use appropriate tie-off to the tower at all times
C. Always wear an approved climbing harness
D. All these choices are correct

35. Which of the following bands has the lowest maximum permissible exposure for RF safety?

A. 3.5 MHz
B. 50 MHz
C. 440 MHz
D. 1296 MHz

Test #2 Answer Key and Explanations

1. C: The Federal Communications Commission (FCC) is the agency that regulates and enforces the rules for the Amateur Radio Service in the United States.

2. B: For a station having a control operator who has been granted a Technician, General, Advanced, or Amateur Extra Class operator license or who holds a CEPT radio-amateur license or IARP of any class. The following frequency bands and segments are authorized to space stations:

(1) The 17 m, 15 m, 12 m, and 10 m bands, 6 mm, 4 mm, 2 mm and 1 mm bands; and

(2) The 7.0-7.1 MHz, 14.00-14.25 MHz, 144-146 MHz, 435-438 MHz, 2400-2450 MHz, 3.40-3.41 GHz, 5.83-5.85 GHz, 10.45-10.50 GHz, and 24.00-24.05 GHz segments.

[97.301, 97.207(c)]

3. D: The person named in an operator/primary station license grant or in a club station license grant is eligible to make application for modification of the license grant, or the renewal thereof, to show a call sign selected by the vanity call sign system. [97.19]

4. B: While an amateur station is broadcasting, one-way transmissions are prohibited. [97.113(b), 97.111(b)]

5. D: Any amateur station may be a space station. A holder of any class operator license may be the control operator of a space station, subject to the privileges of the class of operator license held by the control operator. [97.301, 97.207(c)]

6. C: When using tactical identifiers, a station must transmit its FCC-assigned call sign every 10 minutes. [97.119(a)]

7. A: 146.520 and 446.000 are both simplex frequencies, but only 146.520 is in the 2m band.

8. D: CTCSS is the use of a sub-audible tone transmitted with normal voice audio to open the squelch of the receiver.

9. C: A typical duty of a Net Control Station is to call the net to order and direct communications between stations checking in. The Net Control Station helps maintain order and manages the flow of traffic on the net.

10. B: Vegetation can absorb UHF and microwave signals, significantly weakening them in areas with dense vegetation.

11. A: The orientation of a radio wave's electric field determines its polarization.

12. C: The main benefit to HF radio waves is that they typically bounce off the ionosphere and therefore can have a much greater range. Higher frequencies penetrate the ionosphere and therefore aren't propagated.

13. A: The frequency and power level at which the measurements will be made should be considered when selecting an accessory SWR meter.

14. A: The keypad or the VFO knob can be used to enter the operating frequency on a modern transceiver. A transceiver is a single unit that contains both a radio transmitter and receiver.

15. B: Electrical power is measured in watts. Electrical power is the rate at which electricity can be converted into other types of energy. It is calculated by multiplying the current and the voltage drop. A watt is equal to one joule per second.

16. A: Another way to specify a radio signal frequency of 1,500,000 Hz is 1500 kHz. In the metric system, the prefix kilo- means thousand, so a kilohertz is equal to 1000 hertz. Therefore, 1,500,000 Hz can be converted to kHz by dividing by a thousand: 1,500,000/1000 = 1500 kHz.

17. A: The basic unit of capacitance is the farad. One farad is equal to the capacitance of a capacitor that has an equal and opposite charge of 1 coulomb on each of its plates, and 1 V of potential difference between those plates.

18. A: The formula for calculating voltage in a circuit is voltage equals current multiplied by resistance.

19. C: A potentiometer is often used as an adjustable volume control. A potentiometer is an instrument that measures an unknown voltage by comparing it to a standard voltage.

20. C: A diode allows current to flow in only one direction. A diode is an electronic device that includes two electrodes, an anode and a cathode. These devices can be used to convert an alternating current into a direct current.

21. A: In the figure, component 1 is a resistor.

22. A: A relay may best be described as a switch controlled by an electromagnet. When the device senses a miniscule voltage or current change, it activates a switch or another component in the electric circuit.

23. A: A transceiver is a device that combines the functions of a transmitter and a receiver.

24. A: If a broadcast AM/FM radio is unintentionally receiving amateur radio transmissions, the most likely explanation is that the receiver is unable to reject strong signals outside of the AM/FM bands.

25. B: An antenna analyzer can be used to determine if an antenna is resonant at the desired operating frequency. Antenna analyzers assess the efficiency of antenna systems.

26. B: The correct way to connect a voltmeter to a circuit is in parallel with the circuit. An analog voltmeter, on the other hand, is a moving coil galvanometer that works as a voltmeter after a resistor is installed in series with it.

27. A: The type of modulation most commonly used for VHF packet radio transmission is either FM or PM.

28. B: The result can be that the signal-to-noise ratio received by the repeater can be overpowered by a single signal and thus won't transmit other valid, but weaker signals.

29. B: A directional antenna would be useful for a hidden transmitter hunt. Directional antennas send and receive signals in some directions more than others. Therefore, these antennas are able to ascertain from which direction a signal is being broadcast.

30. B: FT8 is a digital mode capable of low signal-to-noise operation.

31. A: Antenna loading is the term for placing an inductor in series with the antenna so that it reduces the resonant frequency of the antenna.

32. B: The impedance of the most commonly used coaxial cable in typical amateur radio installations is 50 ohms. Impedance is the measure of the total opposition to current flow in an alternating current circuit, composed of reactance and ohmic resistance.

33. D: Electric shock may result in any of the following: cell damage, electrical function disruption in cells, tissue heating, muscle contraction, or burns.

34. D: When climbing an antenna tower it is essential that you have sufficient training on safe tower climbing techniques, use appropriate tie-off to the tower at all times, and wear an approved climbing harness.

35. B: 50 MHz has a lower maximum permissible exposure limit than 3.5 MHz, 440 MHz, or 1296 MHz.

Test #3 Practice Questions

1. What do the FCC rules state regarding the use of a phonetic alphabet for station identification in the Amateur Radio Service?

A. It is required when transmitting emergency messages
B. It is encouraged when using phone emissions
C. It is required when in contact with foreign stations
D. All these choices are correct

2. Which frequency is in the 6-meter amateur band?

A. 49.00 MHz
B. 52.525 MHz
C. 28.50 MHz
D. 222.15 MHz

3. What types of international communications are FCC-licensed amateur radio stations permitted to make?

A. Communications incidental to the purposes of the Amateur Radio Service and remarks of a personal character
B. Communications incidental to conducting business or remarks of a personal nature
C. Only communications incidental to contest exchanges; all other communications are prohibited
D. Any communications that would be permitted by an international broadcast station

4. When is it permissible to transmit messages encoded to obscure their meaning?

A. Only when using a remote station during a contest
B. Only when transmitting certain approved digital codes
C. Only when transmitting control commands to space stations or model craft
D. Never

5. Who must designate the station control operator?

A. The station licensee
B. The FCC
C. The frequency coordinator
D. Any licensed operator

6. When are you required to transmit your assigned call sign?

A. At the beginning of each contact, and every 10 minutes thereafter
B. At least once during each transmission
C. At least every 15 minutes during and at the end of a communication
D. At least every 10 minutes during and at the end of a communication

7. What is a common repeater frequency offset in the 70-centimeter band?

A. Plus or minus 5 MHz
B. Plus or minus 600 kHz
C. Plus or minus 500 kHz
D. Plus or minus 1 MHz

8. Which of the following describes a linked repeater network?

A. A network of repeaters in which signals received by one repeater are transmitted by all the repeaters in the network
B. A single repeater with more than one receiver
C. Multiple repeaters with the same control operator
D. A system of repeaters linked by APRS

9. What technique is used to ensure that voice messages containing unusual words are received correctly?

A. Send the words by voice and Morse code
B. Use the Q-code "QSR" to ask the receiving station to repeat the words back
C. Spell the words using a standard phonetic alphabet
D. All these choices are correct

10. What antenna polarization is normally used for long-distance CW and SSB contacts on the VHF and UHF bands?

A. Right-hand circular
B. Left-hand circular
C. Horizontal
D. Vertical

11. What are the two components of a radio wave?

A. Impedance and reactance
B. Voltage and current
C. Electric and magnetic fields
D. Ionizing and non-ionizing radiation

12. What is one characteristic of VHF signals received via auroral backscatter?

A. They are often received from 10,000 miles or more
B. They are distorted, with a characteristic raspy sound
C. They occur only during winter nighttime hours
D. They are generally strongest when your antenna is aimed west

13. Why are short, heavy-gauge wires used for a transceiver's DC power connection?

A. To minimize voltage drop when transmitting
B. To provide a close match to the power supply output impedance
C. To avoid RF interference
D. To minimize radiative losses in the power cable

14. How is squelch adjusted so that a weak FM signal can be heard?

A. Set the squelch threshold so that receiver output audio is on all the time
B. Turn up the audio level until it overcomes the squelch threshold
C. Turn on the anti-squelch function
D. Enable squelch enhancement

15. What is the term for the flow of electrons in an electric circuit?

A. Voltage
B. Amperes per second
C. Volts per second
D. Current

16. Which is equal to one kilovolt?

A. One one-thousandth of a volt
B. One hundred volts
C. One thousand volts
D. One million volts

17. What describes the ability to store energy in a magnetic field?

A. Admittance
B. Capacitance
C. Resistance
D. Inductance

18. What formula is used to calculate resistance in a circuit?

A. R = E × I
B. R = E / I
C. R = E + I
D. R = E - I

19. What electrical parameter is controlled by a potentiometer?

A. Inductance
B. Resistance
C. Capacitance
D. Field strength

20. Which of these components can be used as an electronic switch?

A. Varistor
B. Potentiometer
C. Transistor
D. Thermistor

21. What is component 2 in the figure below?

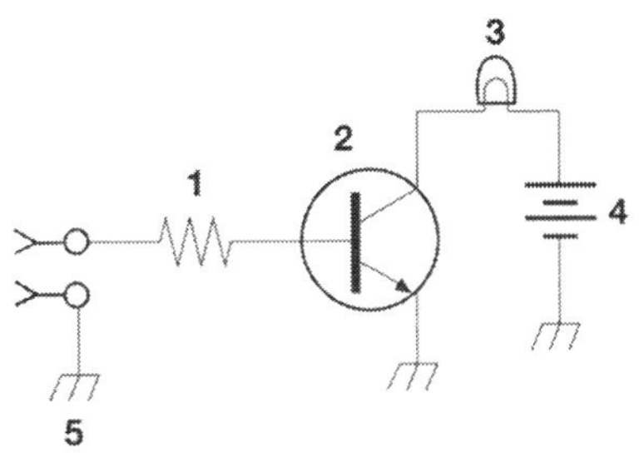

A. Resistor
B. Transistor
C. Indicator lamp
D. Connector

22. Which of the following is a reason to use shielded wire?

A. To decrease the resistance of DC power connections
B. To increase the current carrying capability of the wire
C. To prevent coupling of unwanted signals to or from the wire
D. To reduce receiver overload

23. Which of the following is used to convert a signal from one frequency to another?

A. Phase splitter
B. Mixer
C. Inverter
D. Amplifier

24. Which of the following can cause radio frequency interference?

A. Fundamental overload
B. Harmonics
C. Spurious emissions
D. All these choices are correct

25. What does a typical RF dummy load consist of?

A. A low-voltage power supply and an AC relay
B. A 50-ohm non-inductive resistor mounted on a heat sink
C. A low-voltage power supply and a DC relay
D. A 50-ohm inductive reactance mounted in a shielded enclosure

26. When configured to measure current, how is a multimeter connected to a component?

A. In series
B. In parallel
C. In quadrature
D. In phase

27. Which type of voice mode is often used for long-distance (weak signal) contacts on the VHF and UHF bands?

A. FM
B. DRM
C. SSB
D. PM

28. Which of the following are provided by satellite tracking programs?

A. Maps showing the real-time position of the satellite track over Earth
B. The time, azimuth, and elevation of the start, maximum altitude, and end of a pass
C. The apparent frequency of the satellite transmission, including effects of Doppler shift
D. All these choices are correct

29. What operating activity involves contacting as many stations as possible during a specified period?

A. Simulated emergency exercises
B. Net operations
C. Hidden transmitter hunts
D. Contesting

30. What kind of data can be transmitted by APRS?

A. GPS position data
B. Text messages
C. Weather data
D. All these choices are correct

31. How is the polarization of an antenna described?

A. By the shape of the driven element
B. By the orientation of the electric field
C. By the orientation of the magnetic field
D. By the direction of radiation

32. Why is coaxial cable the most common feed line for amateur radio antenna systems?

A. It is easy to use and requires few special installation considerations
B. It has less loss than any other type of feed line
C. It can handle more power than any other type of feed line
D. It is less expensive than any other type of feed line

33. In the United States, what circuit does black wire insulation indicate in a three-wire 120 V AC cable?

A. Neutral
B. Hot
C. Equipment ground
D. Negative

34. Under what circumstances is it safe to climb a tower without a helper or observer?

A. When no electrical work is being performed
B. When no mechanical work is being performed
C. When the work being done is not more than 20 feet above the ground
D. Never

35. How does the allowable power density for RF safety change if duty cycle changes from 100 percent to 50 percent?

A. It increases by a factor of 3
B. It decreases by 50 percent
C. It increases by a factor of 2
D. There is no adjustment allowed for lower duty cycle

Test #3 Answer Key and Explanations

1. B: By a phone emission in the English language. Use of a phonetic alphabet as an aid for correct station identification is encouraged.

2. B: 52.525 MHz is within the 6-meter band. [97.301(a)]

3. A: FCC-licensed amateur stations are permitted to make international communications incidental to the purposes of the amateur service and remarks of a personal character. [97.117]

4. C: An amateur station is allowed to hide the meaning of a transmitted message with codes or ciphers only when transmitting control commands to space stations or radio control craft. [97.211(b), 97.215(b), 97.113(a)(4)]

5. A: The station licensee must designate the station control operator. [97.103(b)]

6. D: An amateur station is required to transmit its assigned call sign at least every 10 minutes during and at the end of a communication. [97.119(a)]

7. A: A common repeater frequency offset in the 70 cm band is plus or minus 5 MHz.

8. A: A linked repeater network is a network of repeaters that are all repeating the signal received by one repeater of the network.

9. C: To ensure that any proper names and unusual words are copied correctly by the receiving station, it is good practice to spell them out using a standard phonetic alphabet.

10. C: A horizontal antenna polarization is normally used for long-distance weak-signal CW and SSB contacts using the VHF and UHF bands.

11. C: The two components of a radio wave are electric and magnetic fields. This is because a radio wave is an electromagnetic wave.

12. B: A characteristic of VHF signals received via auroral backscatter is that the signals exhibit rapid fluctuations of strength and often sound distorted.

13. A: Heavier gauge wire and short lengths will reduce the resistance as much as possible, thus maintaining sufficient voltage for peak performance.

14. A: Setting the squelch threshold so that receiver output audio is on all the time will help a weak FM signal be heard.

15. D: The flow of electrons in an electric circuit is the current. Current is measured in amperes, also known as amps.

16. C: 1000 V is equal to 1 kV. The metric prefix kilo- means thousand, so a thousand volts is the same as one kilovolt. Therefore, V can be converted to kV by dividing by a thousand: 1000/1000 = 1.

17. D: The ability to store energy in a magnetic field is called inductance. Inductance can also be defined as the characteristic of an electric circuit that enables electromagnetic force to be created in it through a varying magnetic flux.

18. B: The formula for calculating resistance in a circuit is resistance (R) equals voltage (E) divided by current (I). Resistance is the opposition to the passage of current, the result of which is the conversion of electrical energy into heat or another type of energy.

19. B: Resistance is controlled by a potentiometer. A potentiometer will typically have a resistance element connected to the circuit with three terminals. Two of these are attached to the two input voltage conductors of the circuit, and the third is connected to the circuit output. The third terminal is typically able to move across the resistance element, which enables the magnitude of the input voltage to be adjusted.

20. C: A transistor can be used as an electronic switch. It may also be used as a detector.

21. B: In the figure, component 2 is a transistor.

22. C: Shielding a wire is necessary to prevent coupling of unwanted signals to or from the wire.

23. B: A mixer is a device used to convert radio signals from one frequency to another.

24. D: Radio frequency interference may be caused by fundamental overload, harmonics, or spurious emissions. A spurious emission is any radio frequency that was not created or transmitted on purpose.

25. B: A typical RF dummy load consists of a 50-ohm non-inductive resistor mounted on a heat sink. It is designed to absorb transmitter power without radiating a signal.

26. A: An ammeter is usually connected in series with the circuit. In most cases, an ammeter is combined with a voltmeter and ohmmeter in a single instrument.

27. C: The type of voice modulation often used for long-distance or weak signal contacts on the VHF and UHF bands is SSB.

28. D: Satellite tracking programs provide a wealth of information about the path and position of orbiting satellites, as well as their operation.

29. D: Contesting involves contacting as many stations as possible during a specified period of time.

30. D: GPS position data, text messages, and weather data can all be transmitted by APRS.

31. B: The polarization of an antenna is described by the orientation of the electric field.

32. A: Coaxial cable is used more often than any other feed lines for amateur radio antenna systems because it is easy to use and requires few special installation considerations.

33. B: In a three-wire electrical AC plug, the black wire is hot.

34. D: It is never safe to climb a tower without a helper or observer.

35. C: When duty cycle changes from 100 percent to 50 percent, the allowable power density for RF safety increases by a factor of 2.

Test #4 Practice Questions

1. How do you receive official notification of your new license and call sign after passing the exam?

A. Email from the FCC with a link to download the license grant
B. USPS First-Class Mail from the FCC
C. Email from the lead volunteer examiner
D. USPS First-Class Mail from the lead volunteer examiner

2. Which amateur band includes 146.52 MHz?

A. 6 meters
B. 20 meters
C. 70 centimeters
D. 2 meters

3. What may happen if the FCC is unable to reach you by email?

A. Fine and suspension of operator license
B. Revocation of the station license or suspension of the operator license
C. Revocation of access to the license record in the FCC system
D. Nothing; there is no such requirement

4. Under what conditions is an amateur station authorized to transmit music using a phone emission?

A. When incidental to an authorized retransmission of manned spacecraft communications
B. When the music produces no spurious emissions
C. When transmissions are limited to less than three minutes per hour
D. When the music is transmitted above 1280 MHz

5. What determines the transmitting frequency privileges of an amateur station?

A. The frequency authorized by the frequency coordinator
B. The frequencies printed on the license grant
C. The highest class of operator license held by anyone on the premises
D. The class of operator license held by the control operator

6. What language must you use for identification when using a phone emission?

A. Any language recognized by the United Nations
B. Any language recognized by the ITU
C. English
D. English, French, or Spanish

7. What is an appropriate way to call another station on a repeater if you know the other station's call sign?

A. Say "break, break," then the other station's call sign, followed by your call sign
B. Say the station's call sign, then identify with your call sign
C. Say "CQ" three times, then the other station's call sign, followed by your call sign
D. Wait for the station to call CQ, then answer with your call sign

8. Which of the following could be the reason you are unable to access a repeater whose output you can hear?

A. Improper transceiver offset
B. You are using the wrong CTCSS tone
C. You are using the wrong DCS code
D. All these choices are correct

9. What is RACES?

A. An emergency organization combining amateur radio and citizens band operators and frequencies
B. An international radio experimentation society
C. A radio contest held in a short period, sometimes called a "sprint"
D. An FCC Part 97 amateur radio service for civil defense communications during national emergencies

10. What is the effect of antenna cross-polarization over a line-of-sight VHF or UHF path?

A. Modulation sidebands might become inverted
B. Received signal strength is reduced
C. Signals have an echo effect
D. Nothing significant will happen

11. What is the velocity of a radio wave traveling through free space?

A. Speed of light
B. Speed of sound
C. 0.86 times the speed of light
D. 1.86 times the speed of sound

12. Which of the following types of propagation is most commonly associated with occasional strong signals on the 10-, 6-, and 2-meter bands from beyond the radio horizon?

A. Backscatter
B. Sporadic E
C. D region absorption
D. Gray-line propagation

13. How are the audio input and output of a transceiver connected in a station configured to operate using FT8?

A. To a computer running a terminal program and connected to a terminal node controller unit
B. To the audio output and input of a computer running FT8 software
C. To an FT8 conversion unit, a keyboard, and a computer monitor
D. To a computer connected to the FT8converter.com website

14. What does an FM signal sound like when received slightly off frequency?

A. The audio increases in pitch
B. The audio decrease in pitch
C. There is no effect except for reduction in amplitude
D. The audio becomes distorted

15. What term describes the number of times per second that an alternating current makes a complete cycle?

A. Pulse rate
B. Wave number
C. Frequency
D. Wavelength

16. Which is equal to one microvolt?

A. One one-millionth of a volt
B. One million volts
C. One thousand kilovolts
D. One one-thousandth of a volt

17. What is the unit of inductance?

A. Coulomb
B. Farad
C. Henry
D. Ohm

18. What is the resistance of a circuit in which a current of 3 amperes flows when connected to 90 volts?

A. 3 ohms
B. 30 ohms
C. 1/30 ohm
D. 270 ohms

19. What electrical component stores energy in an electric field?

A. Resistor
B. Capacitor
C. Inductor
D. Diode

20. Which of the following components can consist of three regions of semiconductor material?

A. Alternator
B. Transistor
C. Triode
D. Pentode

21. What is component 3 in the figure below?

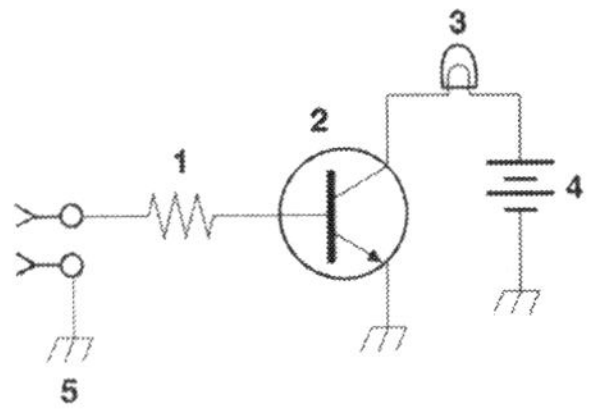

A. Resistor
B. Transistor
C. Lamp
D. Ground symbol

22. Which of the following displays an electrical quantity as a numeric value?

A. Potentiometer
B. Transistor
C. Meter
D. Relay

23. Which term describes the ability of a receiver to discriminate between multiple signals?

A. Discrimination ratio
B. Sensitivity
C. Selectivity
D. Harmonic distortion

24. Which of the following might be the cause of low RF power output from a solid-state transceiver?

A. Poor amplifier noise figure
B. Poor amplifier linearity
C. Low SWR
D. High SWR

25. What reading on an SWR meter indicates a perfect impedance match between the antenna and the feed line?

A. 50:50
B. Zero
C. 1:1
D. Full Scale

26. Which instrument is used to measure electric current?

A. An ohmmeter
B. An electrometer
C. A voltmeter
D. An ammeter

27. Which type of modulation is commonly used for VHF and UHF voice repeaters?

A. AM
B. SSB
C. PSK
D. FM or PM

28. What mode of transmission is commonly used by amateur radio satellites?

A. SSB
B. FM
C. CW/data
D. All these choices are correct

29. Which of the following is good practice when contacting another station in a contest?

A. Signing only the last two letters of your call if there are many other stations calling
B. Contacting the station twice to be sure that you are in his log
C. Sending only the minimum information needed for proper identification and the contest exchange
D. Adding "Please copy" before your exchange

30. What is meant by the term "NTSC?"

A. A digital transmission standard for encrypting data
B. A special mode for satellite uplink
C. An analog fast-scan color TV signal
D. A frame compression scheme for TV signals

31. What is a disadvantage of a handheld radio transceiver's short flexible antenna compared to a full-sized quarter-wave antenna?

A. It has low efficiency
B. It transmits only circularly polarized signals
C. It is more susceptible to receiver desensitization
D. It only works on analog signals, not digital ones

32. What is the major function of an antenna tuner (antenna coupler)?

A. It matches the antenna system impedance to the transceiver's output impedance
B. It helps a receiver automatically tune in weak stations
C. It allows an antenna to be used on both transmit and receive
D. It automatically selects the proper antenna for the frequency band being used

33. What is the purpose of a fuse in an electrical circuit?

A. To prevent power supply ripple from damaging a component
B. To remove power in case of an overload
C. To limit current and prevent shocks
D. All these choices are correct

34. Which of the following is an important safety precaution to observe when putting up an antenna tower?

A. Wear a ground strap connected to your wrist at all times
B. Insulate the base of the tower to avoid lightning strikes
C. Look for and stay clear of any overhead electrical wires
D. All these choices are correct

35. What factors affect the RF exposure of people near an amateur station antenna?

A. Frequency and power level of the RF field
B. Distance from the antenna to a person
C. Radiation pattern of the antenna
D. All these choices are correct

Test #4 Answer Key and Explanations

1. A: Official notification of new licensure and call signs after passing the exam are received by email from the FCC with a link to download the license grant.

2. D: If your station is transmitting on 146.52 MHz, it is using the 2-meter amateur band. [97.301(a)]

3. B: If the FCC is unable to reach you by email, then your station license may be revoked or your operator license may be suspended. [97.23]

4. A: The only time an amateur station is authorized to transmit music is when it is incidental to authorized retransmission of manned spacecraft communications. [97.113(a)(4), 97.113(c)]

5. D: The transmitting privileges of an amateur station are determined by the class of operator license held by the control operator. [97.103(b)]

6. C: English is the acceptable language for station identification use when operating in a phone sub-band. [97.119(b)(2)]

7. B: An appropriate way to call another station on a repeater if you know the other station's call sign is to say the station's call sign and then identify with your call sign.

8. D: There are several common problems that might cause you to be able to hear but not access a repeater even when transmitting with the proper offset. This may occur when the repeater receiver requires audio tone burst, a CTCSS tone, or a DCS tone sequence for access.

9. D: RACES is the Radio Amateur Civil Emergency Service, an FCC Part 97 amateur radio service for civil defense communications during national emergencies. It is a specific emergency communications service defined by FCC rules.

10. B: If the antennas at opposite ends of a VHF or UHF line of sight radio link are not using the same polarization, the signals could be significantly weaker.

11. A: A radio wave travels through free space at the speed of light. This is approximately 300,000,000 meters per second.

12. B: The sporadic E propagation type is most commonly associated with occasional strong signals on the 10, 6, and 2 meter bands.

13. B: In a station configured to operate using FT8, the transceiver audio input and output are connected to the audio input and output of a computer running WSJT-X software.

14. D: An FM signal sounds distorted when it is received slightly off frequency.

15. C: Frequency describes the number of times per second that an alternating current makes a complete cycle.

16. A: 1/1,000,000 of a volt is equal to 1 μV. The metric prefix micro- means millionth, so 1,000,000 microvolts is equal to one volt. Volts can be converted to microvolts by dividing by a million: 1/1,000,000 = 1 μV.

17. C: The basic unit of inductance is the henry.

18. B: The resistance of a circuit in which a current of 3 amperes flows when connected to 90 volts is 30 ohms. Resistance is calculated by dividing voltage by current. In this case, then, the resistance is found 90/3 = 30.

19. B: The capacitor is the electrical component that stores energy in an electric field. Typically, a capacitor is composed of two metallic plates that are separated and insulated from one another by a dielectric.

20. B: A transistor is made of three regions of semiconductor material.

21. C: In the figure, component 3 is a lamp.

22. C: A meter is a device used to display an electrical quantity as a numeric value. Meters can be used to show measurements such as voltage, current, or resistance.

23. C: A receiver's ability to discriminate among multiple radio signals is referred to as its selectivity.

24. D: High SWR might be the cause of low RF power output from a solid-state transceiver.

25. C: An SWR meter reading of 1:1 indicates a perfect impedance match between the antenna and the feed line. This means that all of the transmitted power is being delivered to the load with no reflected power.

26. D: An ammeter is used to measure electric current. Typically, the current will need to be at least a few amperes to be susceptible to measurement by the ammeter.

27. D: The type of modulation most commonly used for VHF and UHF voice repeaters is either FM or PM.

28. D: The three most common transmission modes are SSB, FM, and CW/data.

29. C: Sending only the minimum information needed for proper identification and the contest exchange is good procedure when contacting another station in a radio contest.

30. C: The term NTSC indicates an analog fast scan color TV signal.

31. A: One disadvantage of the antenna supplied with most handheld radio transceivers is it does not transmit or receive as effectively as a full-sized antenna.

32. A: An antenna tuner matches the antenna system impedance to the transceiver's output impedance.

33. B: The purpose of a fuse in an electrical circuit is to interrupt power in case of overload.

34. C: When putting up an antenna tower, it is important to look for and stay clear of any overhead electric wires.

35. D: Several factors affect the RF exposure of people near an amateur station antenna. These factors include the frequency and power level of the RF field, the distance from the antenna to a person, and the radiation pattern of the antenna.

Test #5 Practice Questions

1. What proves that the FCC has issued an operator/primary license grant?

A. A printed copy of the certificate of successful completion of examination
B. An email notification from the NCVEC granting the license
C. The license appears in the FCC ULS database
D. All these choices are correct

2. Which of the following bands include frequencies where Technicians are authorized to use digital modes such as FT8?

A. 10 meters
B. 6 meters
C. 2 meters
D. All these choices are correct

3. Which of the following is a valid Group D call sign format for Technician class?

A. KF1XXX
B. KA1X
C. W1XX
D. All these choices are correct

4. When may amateur radio operators use their stations to notify other amateurs of the availability of equipment for sale or trade?

A. Never
B. When the equipment is not the personal property of either the station licensee, or the control operator, or their close relatives
C. When no profit is made on the sale
D. When selling amateur radio equipment and not on a regular basis

5. What is an amateur station's control point?

A. The location of the station's transmitting antenna
B. The location of the station's transmitting apparatus
C. The location at which the control operator function is performed
D. The mailing address of the station licensee

6. What method of call sign identification is required for a station transmitting phone signals?

A. Send the call sign followed by the indicator RPT
B. Send the call sign using a CW or phone emission
C. Send the call sign followed by the indicator R
D. Send the call sign using only a phone emission

7. How should you respond to a station calling CQ?

A. Transmit "CQ" followed by the other station's call sign
B. Transmit your call sign followed by the other station's call sign
C. Transmit the other station's call sign followed by your call sign
D. Transmit a signal report followed by your call sign

8. Which of the following would cause your FM transmission audio to drop out on voice peaks?

A. Your repeater offset is inverted
B. Your FM deviation is too low
C. You are talking too loudly
D. Your transmit power is too high

9. What does the term "traffic" refer to in net operation?

A. Formal messages exchanged by net stations
B. The number of stations checking in and out of a net
C. Operation by mobile or portable stations
D. A count of the number of activations of the net each month

10. When using a directional antenna, how might your station be able to communicate with a distant repeater if buildings or obstructions are blocking the direct line of sight path?

A. Change from vertical to horizontal polarization
B. Try to find a path that reflects signals to the repeater
C. Try the long path
D. Increase the antenna SWR

11. What is the relationship between wavelength and frequency?

A. Wavelength gets longer as frequency increases
B. Wavelength gets shorter as frequency increases
C. Wavelength is constant at all frequencies
D. Wavelength and frequency increase as path length increases

12. Which of the following effects may allow radio signals to travel beyond obstructions between the transmitting and receiving stations?

A. Knife-edge diffraction
B. Faraday rotation
C. Quantum tunneling
D. Doppler shift

13. Where should an RF power meter be installed?

A. In the feed line, between the transmitter and antenna
B. At the power supply output
C. In parallel with the push-to-talk line and the antenna
D. In the power supply cable, as close as possible to the radio

14. What does the scanning function of an FM transceiver do?

A. Checks incoming signal deviation
B. Prevents interference to nearby repeaters
C. Tunes through a range of frequencies to check for activity
D. Tunes through a range of frequencies to determine the antenna's resonant frequency

15. A difference in which of the following causes electron flow?

A. Voltage
B. Ampere-hours
C. Capacitance
D. Inductance

16. Which is equal to 500 milliwatts?

A. 5 watts
B. 0.5 watts
C. 500,000 watts
D. 500,000,000 watts

17. What is the unit of impedance?

A. Volt
B. Ampere
C. Coulomb
D. Ohm

18. What is the resistance of a circuit for which the applied voltage is 12 volts and the current flow is 1.5 amperes?

A. 18 ohms
B. 0.125 ohms
C. 8 ohms
D. 13.5 ohms

19. What type of electrical component consists of conductive surfaces separated by an insulator?

A. Resistor
B. Potentiometer
C. Oscillator
D. Capacitor

20. What type of transistor has a gate, drain, and source?

A. Varistor
B. Field-effect
C. Hall-effect
D. Bipolar junction

21. What is component 4 in the figure below?

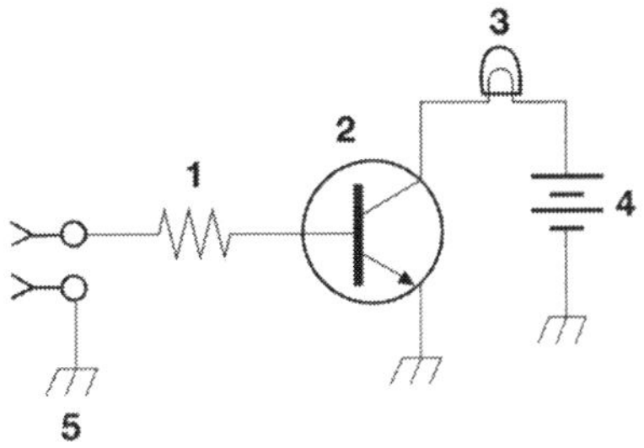

A. Resistor
B. Transistor
C. Ground symbol
D. Battery

22. What type of circuit controls the amount of voltage from a power supply?

A. Regulator
B. Oscillator
C. Filter
D. Phase inverter

23. What is the name of a circuit that generates a signal at a specific frequency?

A. Reactance modulator
B. Phase modulator
C. Low-pass filter
D. Oscillator

24. Which of the following might reduce interference by an amateur station to a non-amateur over-the-air radio receiver?

A. Block the amateur signal with a filter at the antenna input of the affected receiver
B. Block the interfering signal with a filter on the amateur transmitter
C. Switch the transmitter from FM to SSB
D. Switch the transmitter to a narrow-band mode

25. Why do most solid-state transmitters reduce output power as SWR increases beyond a certain level?

A. To protect the RF output amplifier transistors
B. To comply with FCC rules on spectral purity
C. Because power supplies cannot supply enough current at high SWR
D. To lower the SWR on the transmission line

26. Which of the following precautions should be taken when measuring in-circuit resistance with an ohmmeter?

A. Ensure that the applied voltages are correct
B. Ensure that the circuit is not powered
C. Ensure that the circuit is grounded
D. Ensure that the circuit is operating at the correct frequency

27. Which of the following signal types has the narrowest bandwidth?

A. FM voice
B. SSB voice
C. CW
D. Slow-scan TV

28. What is a satellite beacon?

A. The primary transmit antenna on the satellite
B. An indicator light that shows where to point your antenna
C. A reflective surface on the satellite
D. A transmission from a satellite that contains status information

29. What is a grid locator?

A. A letter-number designator assigned to a geographic location
B. A letter-number designator assigned to an azimuth and elevation
C. An instrument for locating faults in power amplifiers
D. An instrument for radio direction finding

30. Which of the following is an application of APRS?

A. Providing real-time tactical digital communications in conjunction with a map showing the locations of stations
B. Automatically showing the number of packets transmitted via PACTOR during a specific time interval
C. Providing voice over internet connection between repeaters
D. Providing information on the number of stations signed into a repeater

31. Which of the following increases the resonant frequency of a dipole antenna?

A. Lengthening it
B. Inserting coils in series with radiating wires
C. Shortening it
D. Adding capacitive loading to the ends of the radiating wires

32. What happens as the frequency of a signal in coaxial cable is increased?

A. The characteristic impedance decreases
B. The loss decreases
C. The characteristic impedance increases
D. The loss increases

33. Why should a 5-ampere fuse never be replaced with a 20-ampere fuse?

A. The larger fuse would be likely to blow because it is rated for higher current
B. The power supply ripple would greatly increase
C. Excessive current could cause a fire
D. Voltage drop in the higher current fuse could result in excessively low voltage to the device

34. What is the purpose of a safety wire through a turnbuckle used to tension guy lines?

A. Secure the guy line if the turnbuckle breaks
B. Prevent loosening of the turnbuckle from vibration
C. Provide a ground path for lightning strikes
D. Provide an ability to measure for proper tensioning

35. Why do exposure limits vary with frequency?

A. Lower frequency RF fields have more energy than higher frequency fields
B. Lower frequency RF fields do not penetrate the human body
C. Higher frequency RF fields are transient in nature
D. The human body absorbs more RF energy at some frequencies than at others

Test #5 Answer Key and Explanations

1. C: When transmitting, each amateur station must have a control operator. The control operator must be a person:

(a) For whom an amateur operator/primary station license grant appears on the ULS consolidated licensee database, or
(b) Who is authorized for alien reciprocal operation by § 97.107 of this part.

2. D: All of the provided choices are correct. [97.301, 97.305]

3. A: Of the choices listed, only the KF1XXX series of call signs is available to technician class amateur operators. The other two series listed are available only to extra class amateur operators.

4. D: Amateur radio operators may use their stations to notify other amateurs of the availability of equipment for sale or trade when the equipment is normally used in an amateur station and such activity is not conducted on a regular basis. [97.113(a)(3)(ii)]

5. C: An amateur station control point is the location at which the control operator function is performed. The control point is the set of locations at which the control operator functions are carried out. [97.3(a)(14)]

6. B: When a station is transmitting phone signals, it should use the call sign identification system of sending the call sign using CW or phone emission. [97.119(b)(2)]

7. C: When responding to a call of CQ, you should transmit the other station's call sign followed by your call sign.

8. C: Speaking too loudly is the most common cause of audio distortions on voice peaks.

9. A: Traffic is a special term for messages on a network.

10. B: When using a directional antenna, a station might be able to access a distant repeater if buildings or obstructions are blocking the direct line of sight path by trying to find a path that reflects signals to the repeater.

11. B: The wavelength of a radio wave gets shorter as the frequency increases. That is, these two characteristics are inversely proportional. If the wavelength grows longer, the frequency will decrease.

12. A: A knife-edge diffraction might allow radio signals to be heard despite obstructions between the transmitting and receiving systems.

13. A: An RF power meter should be installed in the feed line, between the transmitter and antenna.

14. C: The scanning function of an FM transceiver tunes through a range of frequencies to check for activity.

15. A: The electrical term for the electromotive force (EMF) that causes electron flow is voltage. It is useful to think of voltage as the electrical equivalent of water pressure.

16. B: 0.5 W is equivalent to 500 mW. The metric prefix milli- means thousandth, so 1000 mW is the same as 1 W. This conversion can be accomplished by multiplying by a thousand: 0.5 x 1000 = 500.

17. D: The ohm is the unit of impedance.

18. C: The resistance of a circuit with 12 volts applied and 1.5 amperes of current flow is 8 ohms. Resistance is calculated by dividing voltage by current: $12 \div 1.5 = 8$.

19. D: A capacitor consists of two or more conductive surfaces separated by an insulator.

20. B: Field-effect transistors have three pins called gate, drain, and source.

21. D: In the figure, component 4 is a battery.

22. A: A regulator circuit controls the amount of voltage from a power supply.

23. D: An oscillator is a circuit that produces an oscillating signal at a desired frequency.

24. A: Filtering out an unwanted amateur radio signal at the receiver of the impacted device is the best way to reduce or eliminate overload on a TV or radio.

25. A: Solid-state radios will automatically reduce the power output in order to prevent damage, specifically to output amplifier transistors, from an increase in SWR.

26. B: When measuring in-circuit resistance with an ohmmeter, one should ensure that the circuit is not powered.

27. C: CW has a narrower bandwidth than FM voice, SSB voice, or slow-scan TV emission.

28. D: A satellite beacon is a transmission from a space station that contains information about a satellite. These transmissions are used for maintenance and guidance.

29. A: A grid locator is a letter-number designator assigned to a geographic location. The latitude and the longitude of the location are compressed into a brief string of characters, in alternating pairs of letters and numbers.

30. A: The Automatic Packet Reporting System (APRS) is a system designed to provide real-time tactical digital communications along with a map indicating the sources of the signals carrying the communication.

31. C: To make a dipole antenna resonant on a higher frequency, shorten it.

32. D: As the frequency of a signal passing through coaxial cable is increased, the loss generally increases as well.

33. C: It is unwise to install a 20-ampere fuse in the place of a 5-ampere fuse because the excessive current could cause a fire.

34. B: The safety wire prevents the turnbuckle from turning. This will mediate any potential loosening of the turnbuckle due to vibration.

35. D: Exposure limits vary with frequency because the human body absorbs more RF energy at some frequencies than at others.

Test #6 Practice Questions

1. On which of the following HF frequencies can automatically controlled amateur propagation beacons be found?

A. Anywhere in the Technician portion of 10 meters
B. Anywhere, if they are registered with the FCC
C. In a specific portion of each of the HF Technician bands
D. On ten meters, between 28.200 MHz and 28.300 MHz

2. On which HF bands does a Technician class operator have phone privileges?

A. None
B. 10-meter band only
C. 80-meter, 40-meter, 15-meter, and 10-meter bands
D. 30-meter band only

3. Which of the following statements is true about using your amateur radio license when operating aboard a vessel or craft in international waters?

A. Amateur operation is prohibited in international waters
B. You may only operate during maritime emergencies
C. You need special FCC authorization for maritime mobile operation
D. You may operate from a US-documented vessel with the master's permission

4. What, if any, are the restrictions concerning transmission of language that may be considered indecent or obscene?

A. The FCC maintains a list of words that are not permitted to be used on amateur frequencies
B. Any such language is prohibited
C. The International Telecommunications Union (ITU) maintains a list of words that are not permitted to be used on amateur frequencies
D. There is no such prohibition

5. When, other than during an emergency, may a Technician class licensee be the control operator of a station operating in an Amateur Extra Class band segment?

A. At no time
B. When designated as the control operator by an Amateur Extra Class licensee
C. As part of a multi-operator contest team
D. When using a club station whose trustee holds an Amateur Extra Class license

6. Which of the following self-assigned indicators are acceptable when using a phone transmission?

A. KL7CC stroke W3
B. KL7CC slant W3
C. KL7CC slash W3
D. All these choices are correct

7. What is an effective way to seek a call from any phone station when not using a repeater?

A. Sign your call sign once, followed by the words "listening for a call;" if there is no answer, change the frequency and repeat
B. Say "QTC," followed by "this is" and your call sign; if there is no answer, change the frequency and repeat
C. Transmit an unmodulated carrier for approximately 10 seconds, followed by "this is" and your call sign, and pause to listen; repeat as necessary
D. Repeat "CQ" a few times, followed by "this is," and your call sign, then pause to listen; repeat as necessary

8. What type of signaling to a repeater uses two simultaneous audio tones?

A. DTMF
B. CTCSS
C. GMRS
D. D-STAR

9. What is the Amateur Radio Emergency Service (ARES)?

A. A group of licensed amateurs who have voluntarily registered their qualifications and equipment for communications duty in the public service
B. A group of licensed amateurs who are members of the military and who voluntarily agreed to provide message handling services in the case of an emergency
C. A training program that provides licensing courses for those interested in obtaining an amateur license to use during emergencies
D. A training program that certifies amateur operators for membership in the Radio Amateur Civil Emergency Service

10. What is the meaning of the term "picket fencing"?

A. Alternating transmissions during a net operation
B. Rapid flutter on mobile signals due to multipath propagation
C. A type of ground system used with vertical antennas
D. Interference from cable TV in the form of carriers at fixed intervals across the band

11. What is the formula for converting frequency to approximate wavelength in meters?

A. Wavelength in meters equals frequency in hertz multiplied by 300
B. Wavelength in meters equals frequency in hertz divided by 300
C. Wavelength in meters equals frequency in megahertz divided by 300
D. Wavelength in meters equals 300 divided by frequency in megahertz

12. What type of propagation is responsible for allowing over-the-horizon VHF and UHF communications to ranges of approximately 300 miles on a regular basis?

A. Tropospheric ducting
B. D region refraction
C. F2 region refraction
D. Faraday rotation

13. What signals are used in a computer-radio interface for digital mode operation?

A. Receive and transmit mode, status, and location
B. Antenna and RF power
C. Receive audio, transmit audio, and transmitter keying
D. NMEA GPS location and DC power

14. Which of the following controls could be used if the voice pitch of a single-sideband signal returning to your CQ call seems too high or low?

A. The AGC or limiter
B. The bandwidth selection
C. The tone squelch
D. The RIT or Clarifier

15. What is the unit of frequency?

A. Hertz
B. Henry
C. Farad
D. Epicycles per second

16. Which is equal to 3000 milliamperes?

A. 0.003 amperes
B. 0.3 amperes
C. 3,000,000 amperes
D. 3 amperes

17. What is the abbreviation for kilohertz?

A. kHz
B. khz
C. KHz
D. KHZ

18. What is the resistance of a circuit that draws 4 amperes from a 12-volt source?

A. 3 ohms
B. 16 ohms
C. 48 ohms
D. 8 ohms

19. What type of electrical component stores energy in a magnetic field?

A. Resistor
B. Capacitor
C. Inductor
D. Diode

20. How is the cathode lead of a semiconductor diode often marked on the package?

A. With the word "cathode"
B. With a stripe
C. With the letter C
D. With the letter K

21. What is component 6 in the figure below?

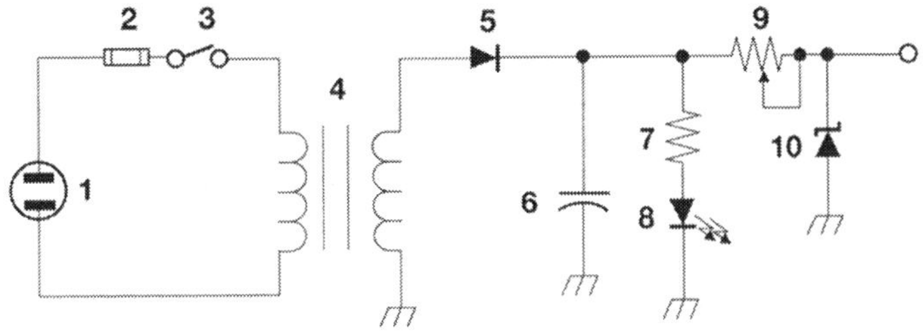

A. Resistor
B. Capacitor
C. Regulator IC
D. Transistor

22. What component changes 120 V AC power to a lower AC voltage for other uses?

A. Variable capacitor
B. Transformer
C. Transistor
D. Diode

23. What device converts the RF input and output of a transceiver to another band?

A. High-pass filter
B. Low-pass filter
C. Transverter
D. Phase converter

24. Which of the following actions should you take if a neighbor tells you that your station's transmissions are interfering with their radio or TV reception?

A. Make sure that your station is functioning properly and that it does not cause interference to your own radio or television when it is tuned to the same channel
B. Immediately turn off your transmitter and contact the nearest FCC office for assistance
C. Install a harmonic doubler on the output of your transmitter and tune it until the interference is eliminated
D. All these choices are correct

25. What does an SWR reading of 4:1 indicate?

A. Loss of -4 dB
B. Good impedance match
C. Gain of +4 dB
D. Impedance mismatch

26. Which of the following can damage a multimeter?

A. Attempting to measure resistance using the voltage setting
B. Failing to connect one of the probes to ground
C. Attempting to measure voltage when using the resistance setting
D. Not allowing it to warm up properly

27. Which sideband is normally used for 10-meter HF, VHF, and UHF single-sideband communications?

A. Upper sideband
B. Lower sideband
C. Suppressed sideband
D. Inverted sideband

28. Which of the following are inputs to a satellite tracking program?

A. The satellite transmitted power
B. The Keplerian elements
C. The last observed time of zero Doppler shift
D. All these choices are correct

29. How is over the air access to Internet Radio Linking Project (IRLP) nodes accomplished?

A. By obtaining a password that is sent via voice to the node
B. By using Dual-Tone Multi-Frequency (DTMF) signals
C. By entering the proper internet password
D. By using Continuous Tone-Coded Squelch System (CTCSS) tone codes

30. What does the abbreviation "PSK" mean?

A. Pulse Shift Keying
B. Phase Shift Keying
C. Packet Sampled Keying
D. Power Sampled Keying

31. Which of the following types of antennas offers the greatest gain?

A. 5/8 wave vertical
B. Isotropic
C. J pole
D. Yagi

32. Which of the following connector types is most suitable as an RF connector for frequencies above 400 MHz?

A. PL-259
B. Type N
C. RS-213
D. DB-25

33. What is a good way to guard against electrical shock at your station?

A. Use three-wire cords and plugs for all AC powered equipment
B. Connect all AC powered station equipment to a common safety ground
C. Ensure all capacitors used for high-voltage DC are fully discharged before working inside equipment
D. All these choices are correct

34. What is the minimum safe distance from a power line to allow when installing an antenna?

A. Add the height of the antenna to the height of the power line and multiply by a factor of 1.5
B. The height of the power line above ground
C. 1/2 wavelength at the operating frequency
D. Enough so that if the antenna falls, no part of it can come within 10 feet of the power wires

35. Which of the following is an acceptable method to determine whether your station complies with FCC RF exposure regulations?

A. By calculation based on FCC OET Bulletin 65
B. By calculation based on computer modeling
C. By measurement of field strength using calibrated equipment
D. All these choices are correct

Test #6 Answer Key and Explanations

1. D: Automatically controlled amateur propagation beacons can be found on ten meters between 28.200 MHz and 28.300 MHz.

2. B: A Technician Class control operator can use 80, 40, 15, and 10m wavelengths, but only the 10m for phone privileges. [97.301(e), 97.305]

3. D: You may operate in international waters from a US-documented vessel with the master's permission. [97.5(a)(2), 97.11(a)]

4. B: As stated by the FCC, no amateur station shall transmit obscene or indecent words or language. [97.113(a)(4)]

5. A: Each class of license allows the licensee to operate only within the allotted frequencies at their level or below. [97.301]

6. D: For identification during a phone transmission, KL7CC stroke W3, KL7CC slant W3, and KL7CC slash W3 would all be acceptable formats for a self-assigned indicator. [97.119(c)]

7. D: Repeating "CQ" a few times, followed by "this is," and your call sign, then pausing to listen is an effective way to seek a call from any phone station when not using a repeater.

8. A: DTMF signals use pairs of audio tones.

9. A: Amateur Radio Emergency Service (ARES) are licensed amateurs who have voluntarily registered their qualifications and equipment for communications duty in the public service

10. B: Picket fencing is a term used to describe the rapid fluttering sound sometimes heard from mobile stations that are moving while transmitting.

11. D: The formula for converting frequency to wavelength in meters is wavelength equals 300 divided by frequency in megahertz.

12. A: Tropospheric ducting is the type of propagation that regularly allows VHF and UHF signals to travel well beyond the horizon for distances of about 300 miles. It is caused by atmospheric conditions, often involving temperature inversions.

13. C: The connections needed are sending and receiving audio in conjunction with transmitter keying in order to allow the computer to trigger the transmitter automatically.

14. D: The RIT or clarifier controls could be used if the voice pitch of a single-sideband signal seems too high or low. Receiver incremental tuning allows the receiver to be tuned up to 3 kilohertz on either side of the transmitter frequency.

15. A: Hertz is the unit of frequency.

16. D: A 3000-milliampere current is equivalent to 3 amperes. The metric prefix milli- means thousandth, so milliamperes can be converted to amperes by dividing by a thousand. An ammeter is an instrument used to measure electric current.

17. A: kHz is the abbreviation for kilohertz.

18. A: The resistance of a circuit that draws 4 amperes from a 12-volt source is 3 ohms. Resistance is calculated by dividing voltage by current: 12/4 = 3.

19. C: An inductor stores energy in a magnetic field. Inductors are typically composed of at least one wire loop. An inductor may be fixed, adjustable, or variable, depending on whether the number of loops is able to be manipulated.

20. B: A semiconductor diode's cathode lead is usually identified with a stripe.

21. B: In the figure, component 6 is a capacitor.

22. B: A transformer is commonly used to change 120V AC house current to a lower AC voltage for other uses.

23. C: A transverter takes the output of a low-powered 28 MHz SSB exciter and produces a 222 MHz output signal. Transverters are a single unit that includes both an upconverter and a downconverter. The transverter enables the transceiver to communicate over a broad range of frequencies.

24. A: If someone tells you that your station's transmissions are interfering with their radio or TV reception, your first step should be to make sure that your station is functioning properly and that it does not cause interference to your own television.

25. D: An SWR reading of 4:1 indicates an impedance mismatch.

26. C: Attempting to measure voltage when using the resistance setting might damage a multimeter. Multimeters are capable of measuring the following values: current, resistance, and the current potential differences of alternating and direct current.

27. A: The upper sideband is normally used for 10 meter HF, VHF, and UHF single-sideband communications.

28. B: The six Keplerian elements precisely define the orbit of a satellite. The six elements are: (1) eccentricity, (2) semi-major axis, (3) inclination, (4) longitude of the ascending node, (5) argument of periapsis, and (6) mean anomaly at epoch.

29. B: Internet Radio Linking Project (IRLP) nodes consist of a radio attached to an internet-connected computer running specialized software. It allows amateur radio users to communicate with other users anywhere in the world near another IRLP node. Users access the node using Dual-Tone Multi-Frequency (DTMF) signals.

30. B: The abbreviation PSK means phase shift keying. Phase shift keying is a strategy for digital modes in the high frequency spectrum. In this system, the components of code are indicated with shifts in the timing relationship between a continuously transmitted tone and a reference tone. This enables the user to transmit a signal in very limited space.

31. D: When compared to 5/8 wave vertical, isotropic, and J pole antennas, Yagi antennas offer the greatest gain.

32. B: A Type N connector is most suitable for frequencies above 400 MHz. Type N connectors are threaded RF connectors used to join coaxial cables. These cables are able to carry microwave frequency signals.

33. D: There are several good ways to guard against electrical shock at your station. Use three-wire cords and plugs for all AC-powered equipment. Connect all AC-powered station equipment to a common safety ground. Also, ensure that capacitors used for high-voltage DC are fully discharged before working inside equipment.

34. D: The minimum safe distance from a power line when installing an antenna is such that, if the antenna falls unexpectedly, no part of it can come closer than 10 feet to the power wires.

35. D: It is acceptable to determine that your station complies with FCC RF exposure regulations by calculation based on FCC OET Bulletin 65, by calculation based on computer modeling, or by measurement of field strength using calibrated equipment.

Test #7 Practice Questions

1. What is the FCC Part 97 definition of a space station?

A. Any satellite orbiting Earth
B. A manned satellite orbiting Earth
C. An amateur station located more than 50 km above Earth's surface
D. An amateur station using amateur radio satellites for relay of signals

2. Which of the following VHF/UHF band segments are limited to CW only?

A. 50.0 MHz to 50.1 MHz and 144.0 MHz to 144.1 MHz
B. 219 MHz to 220 MHz and 420.0 MHz to 420.1 MHz
C. 902.0 MHz to 902.1 MHz
D. All these choices are correct

3. How long before the expiration date may an amateur radio license renewal be requested?

A. 30 days
B. 90 days
C. 6 months
D. 1 year

4. Which of the following is an example of an auxiliary station?

A. A station sending one-way transmissions between a remote repeater receiver and the main repeater transmitter
B. A backup radio for emergency use in case the main station radio fails
C. A station used in the Military Auxiliary Radio System for linking government and amateur stations in a digital network
D. A second station utilized in multi-operator contest stations

5. When the control operator is not the station licensee, who is responsible for the proper operation of the station?

A. All licensed amateurs who are present at the operation
B. Only the station licensee
C. Only the control operator
D. The control operator and the station licensee

6. Which of the following restrictions apply when a non-licensed person speaks to a foreign amateur radio station via a station under the control of an FCC-licensed amateur radio operator?

A. The person must be a U.S. citizen
B. The foreign station must be in a country with which the U.S. has a third-party agreement
C. The licensed control operator must do the station identification
D. All these choices are correct

7. What does the term "repeater offset" mean?

A. The difference between a repeater's transmit and receive frequencies
B. The repeater has a time delay to prevent interference
C. The approximately half-second delay to allow for the "squelch tail" to be removed
D. The fee charged by the repeater owner or club to provide for maintenance

8. How can you join a digital repeater's "talkgroup"?

A. Register with the local Frequency Coordinator
B. Register with the digital repeater control operator
C. Program your radio with the group's ID or code
D. Program your radio with the DTMF tone used by the talkgroup

9. Which of the following is standard practice when you participate in a net?

A. When first responding to the net control station, transmit your call sign, name, and address as in the FCC database
B. Record the time of each of your transmissions
C. Unless you are reporting an emergency, transmit only when directed by the net control station
D. All these choices are correct

10. What weather condition might decrease range at microwave frequencies?

A. High winds
B. Low barometric pressure
C. Precipitation
D. Colder temperatures

11. In addition to frequency, which of the following is used to identify amateur radio bands?

A. The approximate wavelength in meters
B. Traditional letter/number designators
C. Channel numbers
D. All these choices are correct

12. What band is best suited for communicating via meteor scatter?

A. 33 centimeters
B. 6 meters
C. 2 meters
D. 70 centimeters

13. Which of the following is one of the connections required between a computer and a transceiver to operate digital modes?

A. Computer "line out" to transceiver push-to-talk
B. Computer "line in" to transceiver push-to-talk
C. Computer "line in" to transceiver speaker connector
D. Computer "line out" to transceiver speaker connector

14. What is a DMR "code plug"?

A. An adapter cable used to connect a DMR radio to a computer for internet access
B. Configuration data loaded onto your radio to access repeaters and talkgroups
C. An upgrade to DMR programming software provided by the radio manufacturer to accommodate new radio models
D. A Coder-Decoder (CODEC) that converts analog voice data to DMR digital data and vice versa

15. Why are metals generally good conductors of electricity?

A. They have relatively high density
B. They have many free electrons
C. They have many free protons
D. They have relatively high Young's modulus values

16. Which is equal to 3.525 MHz?

A. 0.003525 kHz
B. 35.25 kHz
C. 3525 kHz
D. 3,525,000 kHz

17. What is the abbreviation for megahertz?

A. mHz
B. mHZ
C. Mhz
D. MHz

18. What is the current in a circuit with an applied voltage of 120 volts and a resistance of 80 ohms?

A. 9600 amperes
B. 200 amperes
C. 0.667 amperes
D. 1.5 amperes

19. What electrical component is typically constructed as a coil of wire?

A. Transistor
B. Capacitor
C. Diode
D. Inductor

20. What causes a light-emitting diode (LED) to emit light?

A. Forward current
B. Reverse current
C. Capacitively-coupled RF signal
D. Inductively-coupled RF signal

21. What is component 8 in the figure below?

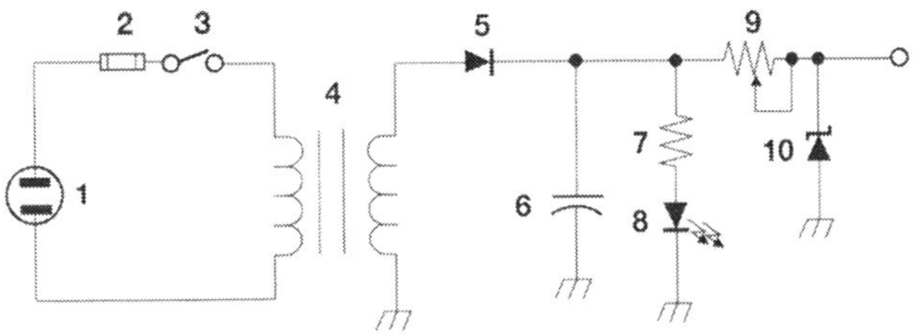

A. Resistor
B. Inductor
C. Regulator IC
D. Light emitting diode

22. Which of the following is commonly used as a visual indicator?

A. LED
B. FET
C. Zener diode
D. All these choices are correct

23. What is the function of a transceiver's PTT input?

A. Input for a key used to send CW
B. Switches transceiver from receive to transmit when grounded
C. Provides a transmit tuning tone when grounded
D. Input for a preamplifier tuning tone

24. Which of the following can reduce interference to a 2-meter band transceiver from a nearby commercial FM station?

A. Installing an RF preamplifier
B. Using double-shielded coaxial cable
C. Installing bypass capacitors on the microphone cable
D. Installing a band-reject filter

25. What happens to power lost in a feed line?

A. It increases the SWR
B. It is radiated as harmonics
C. It is converted into heat
D. It distorts the signal

26. Which of the following measurements are made using a multimeter?

A. Signal strength and noise
B. Impedance and reactance
C. Voltage and resistance
D. All these choices are correct

27. What is one characteristic of single sideband (SSB) compared to FM?

A. SSB signals are easier to tune in correctly
B. SSB signals are less susceptible to interference
C. SSB signals have narrower bandwidth
D. SSB signals are less susceptible to high SWR

28. What is Doppler shift in reference to satellite communications?

A. A change in the satellite orbit
B. A mode where the satellite receives signals on one band and transmits on another
C. An observed change in signal frequency caused by relative motion between the satellite and Earth station
D. A special digital communications mode for some satellites

29. What is Voice Over Internet Protocol (VoIP)?

A. A set of rules specifying how to identify your station when linked over the internet to another station
B. A technique employed to "spot" DX stations via the internet
C. A technique for measuring the modulation quality of a transmitter using remote sites monitored via the internet
D. A method of delivering voice communications over the internet using digital techniques

30. Which of the following describes DMR?

A. A technique for time-multiplexing two digital voice signals on a single 12.5 kHz repeater channel
B. An automatic position tracking mode for FM mobiles communicating through repeaters
C. An automatic computer logging technique for hands-off logging when communicating while operating a vehicle
D. A digital technique for transmitting on two repeater inputs simultaneously for automatic error correction

31. What is a potential drawback of using a handheld VHF transceiver inside a vehicle that lacks an externally mounted antenna?

A. Signal strength is reduced due to the shielding effect of the vehicle
B. The bandwidth of the antenna will decrease, increasing SWR
C. The SWR might decrease, decreasing the signal strength
D. The handheld will overheat due to reflected power in the vehicle

32. Which of the following is true of PL-259 type coax connectors?

A. They are preferred for microwave operation
B. They are watertight
C. They are commonly used at HF and VHF frequencies
D. They are a bayonet-type connector

33. Where should a lightning arrester be installed in a coaxial feed line?

A. At the output connector of a transceiver
B. At the antenna feed point
C. At the AC power service panel
D. On a grounded panel near where feed lines enter the building

34. Which of thc following is an important safety rule to remember when using a crank-up tower?

A. This type of tower must never be painted
B. This type of tower must never be grounded
C. This type of tower must not be climbed unless it is retracted, or mechanical safety locking devices have been installed
D. All these choices are correct

35. What hazard is created by touching an antenna during a transmission?

A. Electrocution
B. RF burn to skin
C. Exposure to ionizing radiation
D. All these choices are correct

Test #7 Answer Key and Explanations

1. C: A space station is an amateur station located more than 50 km above the Earth's surface.

2. A: 50.0 MHz to 50.1 MHz and 144.0 MHz to 144.1 MHz are the only ranges in VHF/UHF that are for CW only. [97.305(a), (c)]

3. B: An amateur radio license renewal may be requested 90 days before the expiration date. [97.21, 1.949]

4. A: A station sending one-way transmissions between a remote repeater receiver and the main repeater transmitter is an example of an auxiliary station. [97.113(d), 97.201(e)]

5. D: When the control operator is not the station licensee, the control operator and the station licensee are both responsible for the proper operation of the station. [97.103(a)]

6. B: An amateur station may transmit messages for a third party to any station within the jurisdiction of any foreign government when transmitting emergency or disaster relief communications and any station within the jurisdiction of any foreign government whose administration has made arrangements with the United States to allow amateur stations to be used for transmitting international communications on behalf of third parties. No station shall transmit messages for a third party to any station within the jurisdiction of any foreign government whose administration has not made such an arrangement. [97.115(a)(2)]

7. A: Repeater offset is the difference between the received and transmitted frequencies of a repeater station.

8. C: Once you have identified the talkgroup you wish to join, all you need is the group's ID or code in order to receive and transmit on their channel.

9. C: To minimize disruptions to an emergency traffic net once you have checked in, do not transmit on the net frequency until asked to do so by the net control station.

10. C: Precipitation can decrease the effective range of microwave frequencies.

11. A: The approximate wavelength (in meters) of radio waves is often used to identify the different frequency bands.

12. B: The 6-meter band is best suited to communicating via meteor scatter.

13. C: Connecting the computer "line in" to transceiver speaker connector enables operation of digital modes.

14. B: A DMR "code plug" contains configuration data loaded onto your radio to access repeaters and talkgroups.

15. B: Metals generally good conductors of electricity because they have many free electrons. This phenomenon of the free electrons is also called the "sea of electrons."

16. C: A frequency of 3.525 MHz is equivalent to 3525 kilohertz. The metric prefix mega- means millions, and the prefix kilo- means thousands. Millions are the same as 10^6, while thousands are equal to 10^3. Megahertz can be converted into kilohertz, then, by multiplying by 10^3, or a thousand: 3.525 MHz x 1000 = 3525.

17. D: MHz is the abbreviation for megahertz.

18. D: The current flow in a circuit with an applied voltage of 120 V and a resistance of 80 ohms is 1.5 amperes. The current flow can be found by dividing the voltage by the resistance: 120/80 = 1.5.

19. D: An inductor is usually composed of a coil of wire.

20. A: An LED is a semiconductor diode that, when a forward current passes through it, produces light in or near the visible range. Each diode is only capable of producing a single primary color, though diodes may be combined to produce any color.

21. D: In the figure, component 8 is a light emitting diode.

22. A: LED is commonly used as a visual indicator.

23. B: The term PTT stands for push-to-talk, a setup in which the transceiver is set to receive when the button is not pushed, and transmit when the button is pushed (when the switch is grounded).

24. D: The purpose of a band reject filter is specifically for blocking specific signals, especially if they are strong, to prevent signal overload.

25. C: The power lost in a feed line is converted into heat. If the cable has cracks or dents, it will be more likely to lose power in this manner. A ham radio operator should always use the best feed line he or she can afford.

26. C: Voltage and resistance measurements are commonly made using a multimeter. The multimeter is sometimes referred to as a volt-ohm meter.

27. C: The primary advantage of single sideband over FM is that SSB signals have narrower bandwidth.

28. C: With regard to satellite communications, Doppler shift is an observed change in signal frequency caused by relative motion between the satellite and the earth station. When the satellite is approaching the earth station, the wavelength will be shorter; as the satellite moves away from the earth station, the wave length will gradually expand.

29. D: Voice-over-Internet Protocol (VoIP) is a method of delivering voice communications over the Internet. It is an integral part of the Internet Radio Linking Project (IRLP).

30. A: DMR is specifically a way to time-multiplex two signals on a single channel.

31. A: One good reason not to use a "rubber duck" antenna inside your car is that signals can be significantly weaker than when it is outside the vehicle.

32. C: PL–259 type coax connectors are commonly used at HF and VHF frequencies.

33. D: When installing a lightning arrester in a coaxial cable feed line, it should be near where the feed lines enter the building and grounded to a common plate which is in turn connected to an external ground.

34. C: An important safety rule to remember when using a crank-up tower is that this type of tower must never be climbed unless it is in the fully retracted position. Crank-up towers are kept stable by a concrete foundation rather than by guy wires.

35. B: If a person accidentally touched your antenna while you were transmitting, they might receive a painful RF burn.

Test #8 Practice Questions

1. Who recommends transmit/receive channels for repeater and auxiliary stations?

A. A Frequency Spectrum Manager appointed by the FCC
B. A Volunteer Frequency Coordinator recognized by local amateurs
C. An FCC Regional Field Office
D. The International Telecommunication Union

2. How are US amateurs restricted in segments of bands where the Amateur Radio Service is secondary?

A. U.S. amateurs may find non-amateur stations in those segments, and must avoid interfering with them
B. U.S. amateurs must give foreign amateur stations priority in those segments
C. International communications are not permitted in those segments
D. Digital transmissions are not permitted in those segments

3. What is the normal term for an FCC-issued amateur radio license?

A. Five years
B. Eight years
C. Ten years
D. Life

4. In which of the following circumstances may the control operator of an amateur station receive compensation for operating that station?

A. When the communication is related to the sale of amateur equipment by the control operator's employer
B. When the communication is a part of classroom instruction at an educational institution
C. When the communication is made to obtain emergency information for a local broadcast station
D. All these choices are correct

5. Which of the following is an example of automatic control?

A. Repeater operation
B. Controlling a station over the internet
C. Using a computer or other device to send CW automatically
D. Using a computer or other device to identify automatically

6. What is the definition of third-party communications?

A. A message from a control operator to another amateur station control operator on behalf of another person
B. Amateur radio communications where three stations are in communications with one another
C. Operation when the transmitting equipment is licensed to a person other than the control operator
D. Temporary authorization for an unlicensed person to transmit on the amateur bands for technical experiments

7. What is the meaning of the procedural signal "CQ"?

A. A shortened form of the term "Contest QSO"
B. "Communication Quality," used to indicate transmitted audio clarity
C. Only the called station should transmit
D. Calling any station

8. Which of the following applies when two stations transmitting on the same frequency interfere with each other?

A. The stations should negotiate continued use of the frequency
B. Both stations should choose another frequency to avoid conflict
C. Whichever station was on the frequency first has preemptive rights to the frequency
D. Use subaudible tones so both stations can share the frequency

9. Which of the following relays messages using email addresses based on amateur callsigns?

A. Winlink
B. FT8
C. PSK31
D. AMTOR

10. What is a likely cause of irregular fading of signals propagated by the ionosphere?

A. Frequency shift due to Faraday rotation
B. Interference from thunderstorms
C. Intermodulation distortion
D. Random combining of signals arriving via different paths

11. What frequency range is referred to as VHF?

A. 30 kHz to 300 kHz
B. 30 MHz to 300 MHz
C. 300 kHz to 3000 kHz
D. 300 MHz to 3000 MHz

12. What causes tropospheric ducting?

A. Discharges of lightning during electrical storms
B. Sunspots and solar flares
C. Updrafts from hurricanes and tornadoes
D. Temperature inversions in the atmosphere

13. Which of the following conductors is preferred for bonding at RF?

A. Copper braid removed from coaxial cable
B. Copper-clad steel wire
C. Twisted-pair cable
D. Flat copper strap

14. What is the advantage of having a choice of receiver filter bandwidths in a multimode transceiver?

A. Permits monitoring several modes simultaneously by selecting a separate filter for each mode
B. Permits noise or interference reduction by selecting a bandwidth matching the mode
C. Increases the number of frequencies that can be stored in memory
D. Increases the amount of offset between receive and transmit frequencies

15. Which of the following is a good electrical insulator?

A. Sea water
B. Glass
C. Stainless steel
D. Graphite

16. Which is equal to 1,000,000 picofarads?

A. 0.001 microfarads
B. 1 microfarad
C. 1000 microfarads
D. 1,000,000,000 microfarads

17. What formula is used to calculate electrical power (P) in a DC circuit?

A. P = I × E
B. P = E / I
C. P = I2 × E
D. P = I / E

18. What is the current through a 100-ohm resistor connected across 200 volts?

A. 20,000 amperes
B. 0.5 amperes
C. 2 amperes
D. 100 amperes

19. What is the function of an SPDT switch?

A. A single circuit is opened or closed
B. Two circuits are opened or closed
C. A single circuit is switched between one of two other circuits
D. Two circuits are each switched between one of two other circuits

20. What does the abbreviation FET stand for?

A. Frequency Emission Transmitter
B. Fast Electron Transistor
C. Free Electron Transmitter
D. Field Effect Transistor

21. What is component 9 in the figure below?

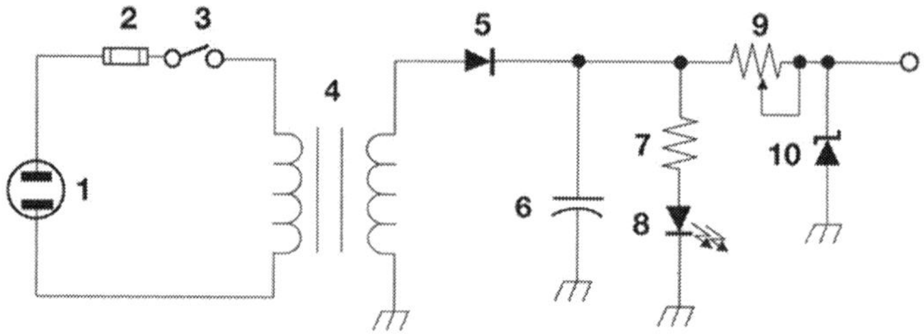

A. Variable capacitor
B. Variable inductor
C. Variable resistor
D. Variable transformer

22. Which of the following is combined with an inductor to make a resonant circuit?

A. Resistor
B. Zener diode
C. Potentiometer
D. Capacitor

23. Which of the following describes combining speech with an RF carrier signal?

A. Impedance matching
B. Oscillation
C. Modulation
D. Low-pass filtering

24. What should you do if something in a neighbor's home is causing harmful interference to your amateur station?

A. Work with your neighbor to identify the offending device
B. Politely inform your neighbor that FCC rules prohibit the use of devices that cause interference
C. Make sure your station meets the standards of good amateur practice
D. All these choices are correct

25. Which instrument can be used to determine SWR?

A. Voltmeter
B. Ohmmeter
C. Iambic pentameter
D. Directional wattmeter

26. Which of the following types of solder should not be used for radio and electronic applications?

A. Acid-core solder
B. Lead-tin solder
C. Rosin-core solder
D. Tin-copper solder

27. What is the approximate bandwidth of a typical single sideband (SSB) voice signal?

A. 1 kHz
B. 3 kHz
C. 6 kHz
D. 15 kHz

28. What does it mean if a satellite is operating in U/V mode?

A. The satellite uplink is in the 15-meter band and the downlink is in the 10 meter band
B. The satellite uplink is in the 70-centimeter band and the downlink is in the 2-meter band
C. The satellite operates using ultraviolet frequencies
D. The satellite frequencies are usually variable

29. What is the Internet Radio Linking Project (IRLP)?

A. A technique to connect amateur radio systems, such as repeaters, via the internet
B. A system for providing access to websites via amateur radio
C. A system for informing amateurs in real time of the frequency of active DX stations
D. A technique for measuring signal strength of an amateur transmitter via the internet

30. Which of the following is included in packet radio transmissions?

A. A checksum that permits error detection
B. A header that contains the call sign of the station to which the information is being sent
C. Automatic repeat request in case of error
D. All these choices are correct

31. Why is a 19-inch-long vertical antenna often used on 2 meters?

A. It has high gain
B. It is a resonant half-wave
C. It is a resonant quarter-wave
D. It has low RF radiation exposure

32. Which of the following is a source of loss in coaxial feed line?

A. Water intrusion into coaxial connectors
B. High SWR
C. Multiple connectors in the line
D. All these choices are correct

33. Where should a fuse or circuit breaker be installed in a 120V AC power circuit?

A. In series with the hot conductor only
B. In series with the hot and neutral conductors
C. In parallel with the hot conductor only
D. In parallel with the hot and neutral conductors

34. Which is a proper grounding method for a tower?

A. A single four-foot ground rod, driven into the ground no more than 12 inches from the base
B. A ferrite-core RF choke connected between the tower and ground
C. A connection between the tower base and a cold-water pipe
D. Separate eight-foot ground rods for each tower leg, bonded to the tower and each other

35. Which of the following actions can reduce exposure to RF radiation?

A. Relocate antennas
B. Relocate the transmitter
C. Increase the duty cycle
D. All these choices are correct

Test #8 Answer Key and Explanations

1. B: A frequency coordinator recommends transmit/receive channels and other parameters for auxiliary and repeater stations. A frequency coordinator is a person or group of people who advise users of the appropriate repeater frequencies, with the intention of diminishing interference between repeaters who are operating in roughly the same frequency in the same geographical area.

2. A: The amateur service is secondary in some portions of the 70 cm band because U.S. amateurs may find non-amateur stations in the bands and must avoid interfering with them. [97.303]

3. C: The normal term for an FCC-issued primary station/operator license grant is 10 years. [97.25]

4. B: The control operator of an amateur station may receive compensation for operating the station when the communication is incidental to classroom instruction at an educational institution. [97.113(a)(3)(iii)]

5. A: A repeater operation may be automatically controlled. [97.3(a)(6), 97.205(d)]

6. A: Third-party communications are any message from the control operator (first party) of an amateur station to another amateur station control operator (second party) on behalf of another person (third party). [97.3(a)(47)]

7. D: The meaning of the procedural signal CQ is "calling any station."

8. A: If two stations transmitting on the same frequency begin to interfere with one another, it is up to the participants to resolve the issue. There is no absolute right to any amateur frequency, so operators are encouraged to remain courteous and work out a solution.

9. A: Winlink relays messages using email addresses based on amateur callsigns.

10. D: The cause of irregular fading of signals from distant stations during times of generally good reception is the random combining of signals arriving via different path lengths.

11. B: The frequency limits of the VHF spectrum are 30 to 300 MHz.

12. D: Tropospheric ducting is caused by temperature inversions in the atmosphere. This is a type of very high frequency propagation that occurs when warm air passes over cold air.

13. D: A flat strap conductor is the best to use for RF grounding. Most ham radio systems have a special terminal for grounding. Copper strap is most often used for grounding.

14. B: The advantage of having a choice of receiver filter bandwidths in a multimode transceiver is that it permits noise or interference reduction by selecting a bandwidth that matches the mode being used. Narrower or wider filters can improve reception depending on the signal type.

15. B: Glass is a good electrical insulator.

16. B: 1,000,000 picofarads is equivalent to 1 microfarad. The metric prefix pico- means trillionths (10^{-12}), while the prefix micro-means millionths (10^{-6}). Picofarads can be converted to microfarads, then, by dividing by 10^6, or one million: 1,000,000/1,000,000 = 1.

17. A: The formula used to calculate electrical power in a DC circuit is power equals voltage multiplied by current.

18. C: The current flowing through a 100-ohm resistor connected across 200 volts is 2 amperes. The current is found by dividing the voltage by the resistance: 200/100 = 2.

19. C: An SPDT switch can toggle a single circuit between one of two other circuits. SPDT stands for Single Pole Double Throw, which means it has a single input and two possible outputs.

20. D: The abbreviation FET stands for Field Effect Transistor. In a field effect transistor, output current is controlled by a variable electric field.

21. C: In the figure, component 9 is a variable resistor.

22. D: A capacitor is used along with an inductor to make a resonant circuit.

23. C: The process of combining speech with an RF carrier signal is referred to as modulation.

24. D: If you believe something in your neighbor's house is causing harmful interference to your amateur station, there are a few things you can or should do. First, make sure that your station is set up properly and that a faulty setup is not causing the interference problems. If that is not successful, talk to your neighbor about the interference problems you are experiencing, and politely let him know about the FCC rules that govern harmful interference. If your neighbor is amenable, work with him to locate the source of the interference and find a solution. Always remain courteous.

25. D: Besides an SWR meter, one could also use a directional wattmeter to determine if a feed line and antenna are properly matched. An SWR (standing wave ratio) meter can evaluate the effectiveness of impedance matching, and can also identify the degree of mismatch between a transmission line and its load. A directional wattmeter is capable of differentiating between the forward and reflected waves.

26. A: Rosin-core solder is best for radio and electronic use. Acid-core should not be used.

27. B: The approximate bandwidth of a single sideband voice signal is 3 kHz.

28. B: When it is said that a satellite is operating in "U/V mode," this means that the satellite uplink is in the 70 cm band and the downlink is in the 2-meter band. In ham radio, modes are particular types of modulation, and modulations are the systems for adding information to a radio signal to be broadcast over the air.

29. A: The IRLP is a project that makes use of the Internet to connect amateur radio stations to one another from all over the world.

30. D: Packet transmissions may include a checksum which permits error detection, a header which contains the call sign of the station to which the information is being sent, and an automatic repeat request in case of error.

31. C: A 19-inch-long vertical antenna is often used on 2 meters because it is a resonant quarter-wave.

32. D: Water intrusion into coaxial connectors, high SWR, and multiple connectors in the line are all sources of loss in a coaxial feed line.

33. A: Home-built equipment that is powered with 120V AC power should always include a fuse or circuit breaker in series with the hot power line to reduce the risk of electrocution hazard.

34. D: Separate eight-foot long ground rods for each tower leg, bonded to the tower and each other, are considered to be a proper grounding method for a tower.

35. A: Amateur operators might relocate antennas to prevent exposure to RF radiation in excess of FCC-supplied limits.

Test #9 Practice Questions

1. Who selects a Frequency Coordinator?

A. The FCC Office of Spectrum Management and Coordination Policy
B. The local chapter of the Office of National Council of Independent Frequency Coordinators
C. Amateur operators in a local or regional area whose stations are eligible to be repeater or auxiliary stations
D. The FCC Regional Field Office

2. Why should you not set your transmit frequency to be exactly at the edge of an amateur band or sub-band?

A. To allow for calibration error in the transmitter frequency display
B. So that modulation sidebands do not extend beyond the band edge
C. To allow for transmitter frequency drift
D. All these choices are correct

3. What is the grace period for renewal if an amateur license expires?

A. Two years
B. Three years
C. Five years
D. Ten years

4. When may amateur stations transmit information in support of broadcasting, program production, or news gathering, assuming no other means is available?

A. When such communications are directly related to the immediate safety of human life or protection of property
B. When broadcasting communications to or from the space shuttle
C. Where non-commercial programming is gathered and supplied exclusively to the National Public Radio network
D. Never

5. Which amateur stations may be remotely controlled?

A. Only repeater stations
B. Only automatically controlled stations
C. Only digital stations
D. Any station

6. What type of amateur station simultaneously retransmits the signal of another amateur station on a different channel or channels?

A. Beacon station
B. Remote control station
C. Repeater station
D. Message forwarding station

7. Which of the following is a customary way to indicate a station is listening on a repeater and looking for a contact?

A. "CQ CQ" followed by the repeater's call sign
B. The station's call sign followed by the word "listening"
C. The repeater's call sign followed by the station's call sign
D. "QSY" followed by your call sign

8. Why are simplex channels designated in the VHF/UHF band plans?

A. So stations within range of each other can communicate without tying up a repeater
B. For contest operation
C. For working DX only
D. So stations with simple transmitters can access the repeater without automated offset

9. Are amateur station control operators ever permitted to operate outside the frequency privileges of their license class?

A. No
B. Yes, but only when part of a FEMA emergency plan
C. Yes, but only when part of a RACES emergency plan
D. Yes, but only in situations involving the immediate safety of human life or protection of property

10. Which of the following results from the fact that signals propagated by the ionosphere are elliptically polarized?

A. Digital modes are unusable
B. Either vertically or horizontally polarized antennas may be used for transmission or reception
C. FM voice is unusable
D. Both the transmitting and receiving antennas must be of the same polarization

11. What frequency range is referred to as UHF?

A. 30 to 300 kHz
B. 30 to 300 MHz
C. 300 to 3000 kHz
D. 300 to 3000 MHz

12. What is generally the best time for long-distance 10-meter band propagation via the F region?

A. From dawn to shortly after sunset during periods of high sunspot activity
B. From shortly after sunset to dawn during periods of high sunspot activity
C. From dawn to shortly after sunset during periods of low sunspot activity
D. From shortly after sunset to dawn during periods of low sunspot activity

13. How can you determine the length of time that equipment can be powered from a battery?

A. Divide the watt-hour rating of the battery by the peak power consumption of the equipment
B. Divide the battery ampere-hour rating by the average current draw of the equipment
C. Multiply the watts per hour consumed by the equipment by the battery power rating
D. Multiply the square of the current rating of the battery by the input resistance of the equipment

14. How is a specific group of stations selected on a DMR digital voice transceiver?

A. By retrieving the frequencies from transceiver memory
B. By enabling the group's CTCSS tone
C. By entering the group's identification code
D. By inserting a five-pin, pre-programmed code plug

15. Which of the following describes alternating current?

A. Current that alternates between a positive direction and zero
B. Current that alternates between a negative direction and zero
C. Current that alternates between positive and negative directions
D. All these answers are correct

16. Which decibel value most closely represents a power increase from 5 watts to 10 watts?

A. 2 dB
B. 3 dB
C. 5 dB
D. 10 dB

17. How much power is delivered by a voltage of 13.8 volts DC and a current of 10 amperes?

A. 138 watts
B. 0.7 watts
C. 23.8 watts
D. 3.8 watts

18. What is the current through a 24-ohm resistor connected across 240 volts?

A. 2400 amperes
B. 0.1 amperes
C. 10 amperes
D. 5760 amperes

Refer to the following for question 19:

19. What type of switch is represented by component 3 in figure T-2?

A. Single-pole single-throw
B. Single-pole double-throw
C. Double-pole single-throw
D. Double-pole double-throw

20. What are the names for the electrodes of a diode?

A. Plus and minus
B. Source and drain
C. Anode and cathode
D. Gate and base

21. What is component 4 in the figure below?

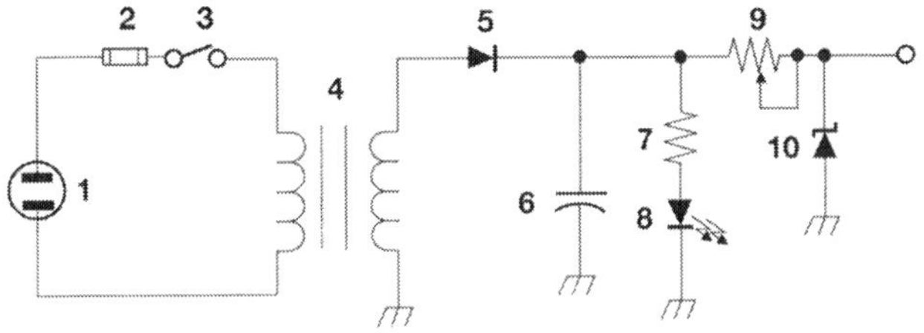

A. Variable inductor
B. Double-pole switch
C. Potentiometer
D. Transformer

22. What is the name of a device that combines several semiconductors and other components into one package?

A. Transducer
B. Multi-pole relay
C. Integrated circuit
D. Transformer

23. What is the function of the switch which selects either SSB or CW-FM on some VHF power amplifiers?

A. Change the mode of the transmitted signal
B. Set the amplifier for proper operation in the selected mode
C. Change the frequency range of the amplifier to operate in the proper segment of the band
D. Reduce the received signal noise

24. What should be the first step to resolve non-fiber optic cable TV interference caused by your amateur radio transmission?

A. Add a low-pass filter to the TV antenna input
B. Add a high-pass filter to the TV antenna input
C. Add a preamplifier to the TV antenna input
D. Be sure all TV feed line coaxial connectors are installed properly

25. Which of the following causes failure of coaxial cables?

A. Moisture contamination
B. Solder flux contamination
C. Rapid fluctuation in transmitter output power
D. Operation at 100% duty cycle for an extended period

26. What is the characteristic appearance of a cold tin-lead solder joint?

A. Dark black spots
B. A bright or shiny surface
C. A rough or lumpy surface
D. A greenish tinge

27. What is the approximate bandwidth of an FM voice signal on VHF repeaters?

A. Less than 500 Hz
B. About 150 kHz
C. Between 10 and 15 kHz
D. Between 50 and 125 kHz

28. What causes spin fading of satellite signals?

A. Circular polarized noise interference radiated from the sun
B. Rotation of the satellite and its antennas
C. Doppler shift of the received signal
D. Interfering signals within the satellite uplink band

29. Which of the following protocols enables an amateur station to transmit through a repeater without using a radio to initiate the transmission?

A. IRLP
B. D-STAR
C. DMR
D. EchoLink

30. What is CW?

A. A type of electromagnetic propagation
B. A digital mode used primarily on 2-meter FM
C. Error correction for digital transmission using code words
D. Another name for a Morse code transmission

31. What is an advantage of a 5/8-wavelength whip antenna for VHF or UHF mobile service compared to a 1/4-wave antenna?

A. It has more gain
B. It radiates at a higher angle
C. It has lower SWR
D. It has lower impedance

32. What can cause erratic changes in SWR?

A. Local thunderstorm
B. Loose connection in the antenna or feed line
C. Over-modulation
D. Overload from a strong local station

33. What should be done to all external ground rods or earth connections?

A. Waterproof them with silicone caulk or electrical tape
B. Keep them as far apart as possible
C. Bond them together with heavy wire or conductive strap
D. Tune them for resonance on the lowest frequency of operation

34. Why should you avoid attaching an antenna to a utility pole?

A. The antenna will not work properly because of induced voltages
B. The antenna may unbalance the power transformer, causing power fluctuations
C. The antenna could contact high-voltage power lines
D. All these choices are correct

35. How can you make sure your station stays in compliance with RF safety regulations?

A. By informing the FCC of any changes made in your station
B. By re-evaluating the station whenever an item in the transmitter or antenna system is changed
C. By making sure your antennas have low SWR
D. By using only Underwriter Laboratories approved transmitting equipment

Test #9 Answer Key and Explanations

1. C: Amateur operators in a local or regional area whose stations are eligible to be auxiliary or repeater stations may select a frequency coordinator. [97.3(a)(22)]

2. D: There are several reasons why an operator should not set the transmit frequency to be exactly at the edge of an amateur band or sub-band. To begin with, this allows for calibration error in the transmitter frequency display. It also ensures that modulation sidebands do not extend beyond the band edge. Finally, it allows for transmitter frequency drift. [97.101(a), 97.301(a-e)]

3. A: The grace period following the expiration of an amateur license, within which the license may be renewed, is two years. [97.21(a)(b)]

4. A: Assuming no other means is available, amateur stations are authorized to transmit signals related to broadcasting, program production, or news gathering only where such communications directly relate to the immediate safety of human life or protection of property. [97.113(5)(b)]

5. D: Any amateur stations may be remotely controlled. [97.109(c)]

6. C: A repeater station is an amateur station that simultaneously retransmits the signal of another amateur station on a different channel or channels. [97.3(a)(40)]

7. B: A call sign followed by the word "listening" is often used in place of CQ to indicate that you are listening on a repeater.

8. A: The main purpose of designating simplex channels is to allow communication directly between sender and receiver and not tie up a repeater.

9. D: Amateur station control operators are permitted to operate outside the frequency privileges of their license class only if necessary in situations involving the immediate safety of human life or protection of property

10. B: Because signals propagated by the ionosphere are elliptically polarized, either vertically or horizontally polarized antennas can be used to transmit and receive them.

11. D: The UHF spectrum is from 300 to 3000 MHz.

12. A: Long-distance 10-meter band propagation via the F region is best conducted between dawn and shortly after sunset during periods of high sunspot activity.

13. B: To determine how long equipment can be powered from a battery, divide the battery's ampere-hour rating by the average current draw of the equipment. This gives the approximate operating time in hours.

14. C: A specific group of stations is selected on a digital voice transceiver by entering the group's identification code.

15. C: As the name implies, alternating current (AC) is current that alternates between positive and negative directions.

16. B: A power increase from 5 to 10 watts creates approximately 3 dB of change. The relationship between power increases and changes in decibel levels is complex. Whenever the power is doubled, however, the decibel level will rise by three. If the power is quadrupled, the decibel level rises by 6. If the power increases by ten times, the decibel level also rises by ten.

17. A: When the voltage is 13.8 volts DC and the current is 10 amperes, the power is 138 watts. Power is calculated by multiplying voltage by current: 13.8 × 10 = 138.

18. C: The current flowing through a 24-ohm resistor connected across 240 volts is 10 amperes. The current is found by dividing the voltage by the resistance: 240/24 = 10.

19. A: A single-pole single-throw switch is represented by component 3 in figure T-2.

20. C: The names of the two electrodes of a diode are the anode and cathode. Electrons leave a system through the anode and return through the cathode.

21. D: In the figure, component 4 is a transformer.

22. C: An integrated circuit combines several semiconductors and other components into one package. All of these semiconductors and other components are either imprinted or etched on a thin slice of a semiconducting material, as for instance silicon.

23. B: Single Side Band/Continuous Wave - FM switch on a VHF amplifier sets the amplifier to the selected mode. The switch provides an easy toggle between settings.

24. D: To remedy any non-fiber optic cable TV interference caused by your amateur radio transmission, ensure that all TV feed line coaxial connectors are installed properly.

25. A: The most common cause of failure in coaxial cables is moisture contamination. Coaxial cable, also known as coax, is composed of a central wire surrounded by hollow tube. The center wire is called the center conductor, and the outer conductor is known as the braid, or shield.

26. C: Cold tin-lead solder joints typically have a rough or lumpy surface.

27. C: The approximate bandwidth of a VHF repeater FM voice signal is between 10 and 15 kHz.

28. B: Spin fading with respect to satellite signals is caused by the rotation of the satellite and its antennas.

29. D: EchoLink enables an amateur station to transmit through a repeater without using a radio to initiate the transmission.

30. D: CW is another name for a Morse code transmission.

31. A: A 5/8-wavelength whip antenna has an advantage over a 1/4-wave antenna because it has more gain.

32. B: A loose connection in an antenna or a feed line might cause erratic changes in SWR readings.

33. C: The best practice for ensuring safety regarding ground rods and earth connections is to bond them together. This can be done with heavy gauge wire or a large conductive strap.

34. C: You should avoid attaching an antenna to a utility pole because the antenna could contact high-voltage power wires.

35. B: A station may ensure compliance by re-evaluating the station whenever an item of equipment is changed.

Test #10 Practice Questions

1. Besides an FCC-issued amateur operator license, what is required to be the control operator of a Radio Amateur Civil Emergency Service (RACES) Station?

A. A written recommendation by the local ARRL Emergency Coordinator
B. Membership in the Amateur Radio Emergency Service (ARES)
C. Certification of current enrollment by a civil defense organization
D. Nothing

2. Where may SSB phone be used in amateur bands above 50 MHz?

A. Only in sub-bands allocated to General class or higher licensees
B. Only on repeaters
C. In at least some segment of all these bands
D. On any band if the power is limited to 25 watts

3. How soon after passing the examination for your first amateur radio license may you transmit on the amateur radio bands?

A. Immediately on receiving your Certificate of Successful Completion of Examination (CSCE)
B. As soon as your operator/station license grant appears on the ARRL website
C. As soon as your operator/station license grant appears in the FCC's license database
D. As soon as you receive your license in the mail from the FCC

4. How does the FCC define broadcasting for the Amateur Radio Service?

A. Two-way transmissions by amateur stations
B. Any transmission made by the licensed station
C. Transmission of messages directed only to amateur operators
D. Transmissions intended for reception by the general public

5. Which of the following is an example of remote control as defined in Part 97?

A. A software defined radio (SDR)
B. Operating the station over the internet
C. Controlling a model aircraft, boat, or car by amateur radio
D. Earth-Moon-Earth (EME) communication

6. Who is accountable if a repeater inadvertently retransmits communications that violate the FCC rules?

A. The control operator of the originating station
B. The control operator of the repeater
C. The owner of the repeater
D. Both the originating station and the repeater owner

7. What is a band plan, beyond the privileges established by the FCC?

A. A voluntary guideline for using different modes or activities within an amateur band
B. A list of operating schedules
C. A list of available net frequencies
D. A plan devised by a club to indicate frequency band usage

8. Which Q signal indicates that you are receiving interference from other stations?

A. QRM
B. QRN
C. QTH
D. QSB

9. What information is contained in the preamble of a formal traffic message?

A. The email address of the originating station
B. The address of the intended recipient
C. The telephone number of the addressee
D. Information needed to track the message

10. What effect does multi-path propagation have on data transmissions?

A. Transmission rates must be increased by a factor equal to the number of separate paths observed
B. Transmission rates must be decreased by a factor equal to the number of separate paths observed
C. No significant changes will occur if the signals are transmitted using FM
D. Error rates are likely to increase

11. What frequency range is referred to as HF?

A. 300 to 3000 MHz
B. 30 to 300 MHz
C. 3 to 30 MHz
D. 300 to 3000 kHz

12. Which of the following bands may provide long-distance communications via the ionosphere's F region during the peak of the sunspot cycle?

A. 6 and 10 meters
B. 23 centimeters
C. 70 centimeters and 1.25 meters
D. All these choices are correct

13. What function does a digital mode hotspot perform for nearby transceivers?

A. Communication with a digital voice or data network
B. FT8 digital communications via AFSK using a smartphone connected to the internet
C. RTTY encoding and decoding without a computer
D. High-speed digital communications for meteor scatter

14. Which of the following receiver filter bandwidths provides the best signal-to-noise ratio for SSB reception?

A. 500 Hz
B. 1000 Hz
C. 2400 Hz
D. 5000 Hz

15. Which term describes the rate at which electrical energy is used?

A. Resistance
B. Current
C. Power
D. Voltage

16. Which decibel value most closely represents a power decrease from 12 watts to 3 watts?

A. -1 dB
B. -3 dB
C. -6 dB
D. -9 dB

17. How much power is delivered by a voltage of 12 volts DC and a current of 2.5 amperes?

A. 4.8 watts
B. 30 watts
C. 14.5 watts
D. 0.208 watts

18. What is the voltage across a 2-ohm resistor if a current of 0.5 amperes flows through it?

A. 1 volt
B. 0.25 volts
C. 2.5 volts
D. 1.5 volts

19. Which of the following battery chemistries is rechargeable?

A. Nickel-metal hydride
B. Lithium-ion
C. Lead-acid
D. All these choices are correct

20. Which of the following can provide power gain?

A. Transformer
B. Transistor
C. Reactor
D. Resistor

21. What is componcnt 3 in the figure below?

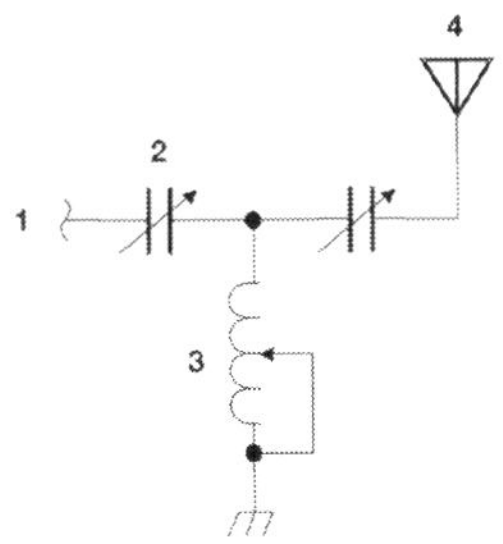

A. Connector
B. Meter
C. Variable capacitor
D. Variable inductor

22. What is the function of component 2 in the figure below?

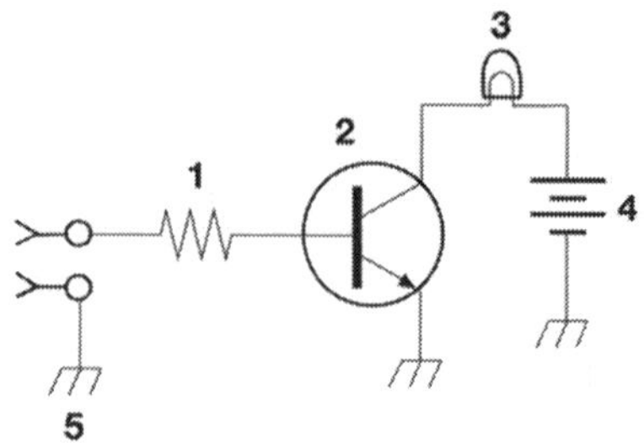

A. Give off light when current flows through it
B. Supply electrical energy
C. Control the flow of current
D. Convert electrical energy into radio waves

23. What can be added to the output of a transceiver to increase the transmitted output power?

A. A potentiometer
B. An RF power amplifier
C. An impedance multiplier
D. All these choices are correct

24. What might be a problem if you receive a report that your audio signal through an FM repeater is distorted or unintelligible?

A. Your transmitter is slightly off frequency
B. You are speaking too loudly or too close to the microphone
C. You are in a bad location
D. All these choices are correct

25. Why should the outer jacket of coaxial cable be resistant to ultraviolet light?

A. Ultraviolet light can increase the resistance of the conductors
B. Ultraviolet light can increase losses in the cable's jacket
C. Ultraviolet and RF signals can mix, causing interference
D. Ultraviolet light can damage the jacket and allow water to enter the cable

26. What reading indicates that an ohmmeter is connected across a large, discharged capacitor?

A. Increasing resistance with time
B. Decreasing resistance with time
C. Steady full-scale reading
D. Alternating between open and short circuit

27. What is the approximate bandwidth of AM fast-scan TV transmissions?

A. More than 10 MHz
B. About 6 MHz
C. About 3 MHz
D. About 1 MHz

28. What does the term LEO mean in reference to communication satellites?

A. Low Energy Orbit, which conserves battery power
B. Low Elevation Orbit, which appears close to the horizon from the earth station
C. Low Equilibrium Orbit, which has a slightly unstable period
D. Low Earth Orbit, which has a period of around 100 minutes

29. What is required before using the EchoLink system?

A. Complete the required EchoLink training
B. Purchase a license to use the EchoLink software
C. Register your call sign and provide proof of license
D. At least a General Class license

30. Which of the following operating activities is supported by digital mode software in the WSJT-X software suite?

A. Earth-Moon-Earth
B. Weak signal propagation beacons
C. Meteor scatter
D. All these choices are correct

31. In which direction does a half-wave dipole antenna radiate the strongest signal?

A. Equally in all directions
B. Off the ends of the antenna
C. In the direction of the feed line
D. Broadside to the antenna

32. What is the electrical difference between RG-58 and RG-213 coaxial cable?

A. There is no significant difference between the two types
B. RG-58 cable has two shields
C. RG-213 cable has less loss at a given frequency
D. RG-58 cable can handle higher power levels

33. What hazard exists when rapidly charging or discharging an unprotected battery?

A. Overheating or out-gassing
B. Excess output ripple
C. Electric shock
D. Overvoltage

34. Which of the following is true when installing grounding conductors used for lightning protection?

A. Use only non-insulated wire
B. Wires must be carefully routed with precise right-angle bends
C. Sharp bends must be avoided
D. Common grounds must be avoided

35. Why is duty cycle one of the factors used to determine safe RF radiation exposure levels?

A. It affects the average exposure to radiation
B. It affects the peak exposure to radiation
C. It takes into account the antenna feed line loss
D. It takes into account the thermal effects of the final amplifier

Test #10 Answer Key and Explanations

1. C: Certification of current enrollment by a civil defense organization is required, in addition to an FCC-issued amateur operator license, to be the control operator of a RACES station. [97.407(a)]

2. C: There is some part of all the amateur bands above 50 MHz that is available for SSB. [97.305(c)]

3. C: After you pass the examination required for your first amateur radio license, you may operate a transmitter on an amateur service frequency as soon as your name and call sign appear in the FCC's ULS database. [97.5a]

4. D: In the FCC rules for the amateur services, broadcasting refers to transmissions intended for reception by the general public. Broadcasting refers to transmissions that are meant to be received by the general public, either on relay or directly. [97.3(a)(10)]

5. B: Operating the station over the internet is an example of remote control. [97.3(a)(39)]

6. A: If a repeater inadvertently retransmits communications that violate the FCC rules, the control operator of the originating station is accountable. [97.205(g)]

7. A: A band plan, beyond the privileges established by the FCC, is a voluntary guideline for using different modes or activities within an amateur band.

8. A: The Q signal used to indicate that you are receiving interference from other stations is QRM. Q signals are useful methods for accelerating communication. A Q signal may be issued as a question or as a response. For example, QRG may mean "What is my exact frequency?" or "Your exact frequency is."

9. D: The preamble in a formal traffic message is information needed to track the message as it passes through the amateur radio traffic handling system. The preamble is the first part of a radiogram. It identifies the source and some basic information about the message.

10. D: If the data signals propagate over multiple paths, error rates are likely to increase.

11. C: The frequency range of 3 to 30 MHz is referred to as HF.

12. A: During periods of high sunspot activity, the 6 meter and 10 meter bands may be used for long-distance communication.

13. A: Communication using digital voice or data systems via the internet is performed with a transceiver and a digital mode hot spot.

14. C: 2400 Hz is an appropriate receive filter to select in order to provide the best signal-to-noise ratio for SSB reception.

15. C: The rate at which electrical energy is used is power.

16. C: A power decrease from 12 watts to 3 watts would result in approximately 6 dB of change. Just as a quadrupling of power would result in an extra 6 dB, so will a quartering of power cause the decibel level to drop by 6.

17. B: When the applied voltage is 12 volts DC and the current is 2.5 amperes, 30 watts of power are generated. The power can be calculated by multiplying the voltage by the current: 12 x 2.5 = 30.

18. A: The voltage across a 2-ohm resistor if a current of 0.5 amperes flows through it is 1 volt. The voltage can be calculated by multiplying the current by the resistance: 2 x 0.5 = 1.

19. D: There are many different types of batteries that may be recharged, including nickel-metal hydride (NiMH), lithium-ion (Li-ion), and lead-acid.

20. B: Transformers don't provide a gain. From the options provided, only a transistor provides a gain.

21. D: In the figure, component 3 is a variable inductor.

22. C: In the figure, the function of component 2 is to control the flow of current.

23. B: An RF power amplifier increases the low-power output from a handheld transceiver.

24. D: If you receive a report that your audio signal through the repeater is distorted or unintelligible, this may be because your transmitter is slightly off frequency, your batteries are running low, or you're in a bad location.

25. D: The outer jacket of a coaxial cable should be resistant to ultraviolet light because this light can damage the jacket and allow water to enter the cable.

26. A: When an ohmmeter is connected across a large, discharged capacitor it will show increasing resistance over time.

27. B: The typical bandwidth of AM fast-scan TV transmissions on the 70 cm band is about 6 MHz.

28. D: The initials LEO indicate that the satellite is in a Low Earth Orbit. This means that the path of the satellite is less than 1000 miles above the surface of the Earth.

29. C: You must provide your call sign and proof of license since this type of use is a type of remote control of a station.

30. D: Earth-Moon-Earth, weak signal propagation beacons, and meteor scatter are all well suited to the short transmissions of WSJT.

31. D: The radiation is strongest from a half-wave dipole antenna in free space broadside to the antenna.

32. C: An RG-213 cable has less loss at a given frequency than an RG-58 cable.

33. A: If a battery is charged or discharged too quickly, the battery could overheat and give off flammable gas, or explode.

34. C: Sharp bends must be avoided in the grounding conductors used for lightning protection.

35. A: Duty cycle is one of the factors used to determine safe RF radiation exposure levels because it affects the average exposure of people to radiation. If the duty cycle is lower, then there is less RF radiation exposure with no change in peak envelope power output.

Online Resources

Due to our efforts to try to keep this book to a manageable length, we've created a link that will give you access to all of your online resources:

mometrix.com/resources719/hamradiotech-33736

It's Your Moment, Let's Celebrate It!

Share your story @mometrixtestpreparation

Made in United States
Orlando, FL
10 September 2026

87579354R00098